THE

WOODS AND BY-WAYS

OF

NEW ENGLAND.

BY

WILSON FLAGG,

AUTHOR OF "STUDIES IN FIELD AND FOREST."

With Illustrations.

The temples of the gods made desolate,
They leave the earth to curses born of art;
Degenerate man resumes the bow and quiver,
And beauty sleeps until another dawn.

BOSTON:
JAMES R. OSGOOD AND COMPANY,
LATE TICKNOR & FIELDS, AND FIELDS, OSGOOD, & CO.
1872.

UNIVERSITY PRESS: WELCH, BIGELOW, & CO.,
CAMBRIDGE.

DEDICATORY EPISTLE.

To DANIEL RICKETSON, Esq.,

Author of "The Autumn Sheaf," etc.

My dear Ricketson: —

Soon after my "Studies in Field and Forest" appeared, you mentioned, as one of the faults of the book, that the author is not sufficiently identified with it, and so rarely alludes to himself or his adventures that it wants the interest which a little egotism would impart to it. I observe also that Thoreau, in one of his "Letters," complains of my lack of enthusiasm. As Thoreau and I never met, he must have formed this opinion from my writings; but those who know me and my habits would say that my life has been too retired for that sort of personal adventure which inspires enthusiasm, or creates a necessity for making self one of the subjects of discourse. My life has been passed with my family in almost entire seclusion, hardly interrupted by a small circle of friends and kinsmen, who, being engaged in trade, have not been my companions; for men of letters and commercial men, how much so ever they may hold each other in mutual esteem, are seldom intimates. And as I have had no social intercourse with any person who is distinguished in science, literature, the fine arts, or by wealth, politics, or civil position, I have lived almost alone in the world. I have devoted my social hours exclusively to my own family, and having had access, until my late domiciliation in Cambridge, to but few books, I have studied Nature more than the library, employing my time in observing her aspects and interpreting her problems, more than in reading or hearing the observations of others.

Few men save those who from religious motives have renounced the world have lived so little in communication with it as I have. I am not a member of any society or club, of any church or institution, trade, profession, or organization. Though once a student of Harvard College, I am not a graduate; and though in my early manhood for many years a contributor to the political press, I have never been an editor nor a politician. I have lived entirely without honors, and have never rejected any. And if, possibly, I have on any occasion manifested an appreciable amount of boldness or independence in speaking my thoughts and avowing my opinions, any such eccentricity may be attributed to this circumstance; for every honor a man receives from the community is a fetter upon his freedom of speech and action. I have not been drawn into society by a taste for its amusements or its vices; I have not joined the crowd either of its saints or its sinners; I have pursued my tasks alone, except as I have read and conversed with my wife and children. She and they have been the only companions of my studies and recreations during all the prime of my life. But, perhaps from this cause alone, I have been very happy. The study of nature and my domestic avocations have yielded me a full harvest of pleasures, though it was barren of honors.

When you read this volume, you will discover, if you open it as a work of technical exactness in its descriptions of natural objects, that it has no such merit. Though I have probably passed more time in the woods than any man who is not a woodcutter by trade, I have not been a collector of specimens, nor a dissector of birds and flowers, nor a measurer of trees, nor a hammerer of rocks. I know the value of this kind of research, but my observations are of a different character. I distinguish the objects of nature as I distinguish my friends by physiognomical marks. My book differs from learned works as Lavater's "Physiognomy" differs from Cheselden's "Anatomy," or as a lover's description of his lady's hand would differ from Bell's anatomical description of it. I mention these things, not with any vulgar depreciation of technical science, but that the reader may not seek in this volume for matters which it does not contain.

In describing the aspects of nature, I have selected such views as afford me the most pleasure, endeavoring by my manner of presenting them to inspire the reader with the same agreeable sensations. I have aimed, not so much to make a graphic picture of any scene from which a painter might with his brush or pencil obtain a copy on canvas, as, on the other hand, to make the reader feel as he would in the presence of it. I have also confined my descriptions to ordinary scenes. These alone have been my study. The objects that meet our view in our walks outside of any village in the country, the beauty of a plain cottage and its picturesque inmates, with their baskets of whortleberries and their bundles of dried herbs, and the common trees and shrubs of the forest and the wayside, form the subjects of my essays. From them I have studied the oracles of nature, and in these pages I have given their interpretations as I understand them.

Some of my friends have asked me why I selected so hackneyed a topic as nature, whose beauties and whose phases have been so often described that every sentence one may write on this subject can hardly be anything more than the repetition of some platitude. I reply that I have described these things because I am familiar with them, and may treat of them without offending popular prejudices, as I might if I were to discourse upon ethics or politics. But the subjects I have chosen are not so hackneyed as many suppose them to be. Popular writers on Nature's aspects have generally been tourists or landscape gardeners; and her grander scenes have been selected by one class, and artificial or dressed landscape by the other. These matters, as the reader will soon discover, have no part in my descriptions. I ought to allude also to the writers on landscape painting, who, with all their professed admiration of Nature, always place her in subordination to art.

With regard to the style of these essays, I will only say that it has been my principal aim to express my thoughts with clearness and simplicity; and as metaphors, except in rare instances, tend to obscurity, I have not sought for them as embellishments. Though a certain vagueness of description is often favorable to our purpose if we would only excite sensations, precision is the first point to be attained when we would convey to the reader's mind a philosophic truth. I have not studied to express my thoughts by any peculiarity of language, but by the use of simple and common terms to render them lucid and interesting.

In you, my dear sir, I have in the autumn of my life met with a friend from whom I have learned to view nature in a new variety of aspects; to you I would respectfully dedicate this volume, and take this opportunity to acknowledge the pleasure I have derived from your friendship, and to assure you how much I feel honored by it.

WILSON FLAGG.

TO THE READER.

I HAVE written this volume not with any desire to stay the progress of those improvements which are necessary to the wants of an increasing population. We are carried along by an irresistible current, and any effort to stay it would be a striving against fate. But as a river may to a certain extent be directed in its course, though it cannot be stopped, in like manner may the progress of the civilized arts be modified by a common intelligence, so as not to destroy the land whose population they sustain. My object is to inspire my readers with a love of nature and simplicity of life, confident that the great fallacy of the present age is that of mistaking the increase of the national wealth for the advancement of civilization. Our peril lies in the speed with which every work goes forward, rendering us liable, in our frantic efforts to grasp certain objects of immediate value, to leave ruin and desolation in our track which will render worthless all the desirable objects we have attained. In this work I have discussed its several points chiefly with reference to our material welfare. The ethical part of the subject I have treated more fully in an unpublished volume entitled "The Progress and Perils of Civilization in America."

DOMESTIC SCENERY OF NEW ENGLAND.

When journeying in New England you cannot fail to be charmed with those old roads that pass through the ruder parts of the early settlements which have not been changed by the improvements that follow any sudden increase of commercial prosperity. Many of them, which at first were highways, are at present only by-roads to some little hamlet, situated apart from the great thoroughfares of commerce, and retaining the simplicity of a former era. It is delightful to enter by chance upon one of these old roads, when it will carry you half a day's journey on foot, without the intrusion upon your sight of a steam-factory or a railroad station. Some of these ways are not traversed enough to obliterate the two rows of grass in the middle of the road, so suggestive of quiet and homely retirement. The farm-houses that meet your sight are among the few remaining examples of the simple style of building that prevailed here during the last century. These and the objects connected with them form the most interesting and representative scenery of New England, and I mark and admire them as distinguishing this country from all the rest of the world.

Some people look upon these scenes as points where progress and civilization are at a stand, and turn away from them with displeasure. But there is another view that is more rational and nearer the truth. These objects, though not borne on the great tide of civilization, are some of its most beneficent results. If you watch a river flowing impetuously over plains and through valleys, you may suppose its moving mass of waters to represent the great highways and thoroughfares of commerce, and to emblem the progress and enterprise of man. But the beauties of the river are little shallows of still water covered with aquatic flowers, and green masses of shrubbery that afford a harbor to the singing-birds. These quiet and flowery inlets, fed by the stream, but not joining in its motion, represent the rural hamlets described in this essay. They are nurtured by the arts and refined by the culture, but not corrupted by the vices, nor disturbed by the ambition, of the great world. Were it not for the river's moving mass of waters these quiet inlets of beauty could not exist; and without this impetuous tide of commerce and the arts, these remote hamlets would not have attained civilization. But as the world moves onward, its learning and culture, its virtue and happiness, turn aside and linger in these rural retreats.

When passing over the old roads of New England, you must take heed that you are not led out of their course by some new and shorter cut. The road that winds around the hill or the meadow is the path you must follow. On the improved road you will see gravel and loam, nice new houses and painted fences, with stiff spruces in their enclosures, and perhaps a formal clipped hedge-row in front. The old road is bordered with wild shrubbery, groups of trees of bold and irregular growth, and here and there a solitary standard, always charmingly out of place. There is no sameness in your journey. You will hardly travel a furlong through the woods before you arrive at an open space that exposes to view some beautiful meadow, lying several feet below the level of the winding road. A small river flows in an irregular course along the interval, often passing out of sight behind some wooded eminence, then reappearing, its surface radiant with purple and amethyst, now smooth as a mirror, then gleaming and sparkling from a thousand rippling waves. Nothing can surpass the grouping of the woods in these natural openings, enlivened with an occasional farm-house, its barns and sheds and peaceful flocks, and revealing in the distance the church-spire of a neighboring hamlet. As the trees consist chiefly of maple, ash, and tupelo, with a few oaks, and a border growth of cornels, viburnums, and whortleberry-bushes, you should visit one of these places to see the most beautiful display of autumnal wood-scenery.

Whatever course you may take, you will arrive occasionally at a railroad station; but the new village suddenly built upon any such point is without peculiar attractions. Some of the houses are models of elegance, but they are like all others in the busy world. These new villages are the cosmopolitan parts of New England, displaying models of perfection in ornate art, and exposing to your observation only what may be seen in every new city. Their scenery is not what the picturesque eye is looking for, and fails to represent the special features of this part of the country. The glare, the art, the taste, fashion, and ostentation apparent in the new houses in these new places are ornamental patches upon the landscape, and are not peculiar to New England.

The old roads in the Northeastern States, except the turnpikes, were never "laid out." They are but the widening of paths made by pedestrians going from one house to another, or of the cartways of the pioneer farmer and woodman. They are generally somewhat elevated, unless they are carried over a plain. They are situated a little above the base of the hills and eminences which they encircle, to avoid the wet grounds and the entanglements of vines and shrubbery that crowd the borders of all the lowlands. All along the course of these primitive roads are constantly rising to view plain farmhouses, with their barns and barnyards, their wells with cross-poles, their woodsheds, their workshops, and their few domestic animals. Many of these houses were originally painted red, with white facings. Some were without paint, except their white borders, neatly contrasted with the dark stone-color of the wooden walls. The houses are generally set back a few rods from the highway and shaded by elms. They are not enclosed, and the wide slope between the house and the road is grazed by the farmer's cattle.

In the rear of the house is a cartway leading between two irregular rows of hickories, oaks, butternuts, and wild-cherry-trees, — the gratuitous product of nature and chance. The predominance of nut-bearing trees in these lanes was caused by the squirrels that harbor in the loose stone-walls and hoard their surplus of nuts by planting them under the shrubbery in the borders. This path leads to a wood-lot, and is often continued through the forest, making one of those green avenues without which we could not realize half the attractions of a wood. Sometimes the farm-house is located a good distance from the road, and is approached by a lane gliding through a half-wooded meadow, and bordered with Lombardy poplars. In the course of your journey you may discover a house and farm enclosed on all sides by the forest, when it seems a little paradise. But our country-houses generally stand near the road, or distant from it only a few paces.

The New England farmer is a hard-working man; for his land is neither very deep nor productive, and with the help of his sons, or perhaps one hired man, he performs all the labor upon it. He gains a small revenue by selling the products of the farm; but if this were his only resource, his lot would be hard. Adjoining the house, or not far from it, usually a little nearer the road, is a small building with a single door and three or four windows, used for a workshop. When his harvest is gathered, he lays aside the ploughshare and the reaping-hook, and takes up the lapstone for his winter's occupation. The farm supplies his household with domestic products, but his pecuniary gains come chiefly from his labors as a shoemaker.

All my life have I admired these little picturesque workshops, when traversing the old roads that lead from one village to another. They are perfectly plain and simple in their style, but as neat as they are unadorned, and beautiful from their expression of the quiet and industrious habits of the people who occupy them. There are no objects in village scenery that so pleasantly harmonize with the cheerful scenes of nature as the plain cottages on these roads and their little adjacent shoemaker's shops. Nothing in the world could so plainly express the union of comfort, freedom, and independence. In Europe no such objects are to be seen. There the houses of the peasantry are not scattered in this charmingly picturesque manner over the land. They are huddled together in cantonments, like the Irish houses in the suburbs of our cities, seldom leaving any space for a garden, and rendering neatness and cleanliness impossible.

The plain and economical system of agriculture still prevailing in many parts of the country, where the only changes that have been adopted are improvements in tillage and implements, has left the face of nature undespoiled of its native embroidery by the vandalism of taste. Here the country is still charming to every philanthropist. We may walk, in many parts, over a distance of several miles of such landscape, interspersed with hundreds of plain houses and their workshops, as beautiful as they are plain and simple, and as picturesque as the wild vines that trail over their fences. But these charming scenes are rapidly disappearing, and in the same ratio is village landscape growing ostentatious and insipid, showing forth the vanity of the owners and artists, and concealing the occupations and all the interesting habits of the villagers under a vapid counterfeit of the fashions of cities.

There are few things more agreeable in village scenery than the evidences of independent labor as distinguished from associated labor under an overseer. Hence the beauty of those little shoemaker's shops, formerly so numerous in the country, and, on the other hand, the gloomy appearance of large buildings for manufactures. Even if there were proof that the operatives in the employ of a capitalist are as comfortable, as thrifty, and as happy as if they were independent workmen, we still associate subordinate labor with the ambitious striving of a few at the expense of the many. A factory village, where the homes of those who labor are in large tenement-blocks, and the only houses outside of the village are the ornate residences of masters and superintendents, is vapid and uninteresting. Farm labor is rapidly losing its independent character in a similar way, by the gradual absorption of agricultural property into the hands of wealthy mortgagees, and the conversion of independent farm laborers into menials. Therefore do we with the more satisfaction recur to these vestiges of New England simplicity, where the farmer is still a yeoman, and look with delight upon the single workshops in many parts of our land, still scattered among the neat and humble cottages, —a smithy in the heart of a little settlement, a saw-mill turned by a brook, and other buildings devoted to independent labor.

You will seldom pass a country village without seeing a graveyard in its vicinity; but the old grounds in which slate has not been displaced by white marble are the only picturesque objects of this kind. Our ancestors selected as their burial-place a quiet spot not far from the village, and did not plant it with trees because it was surrounded by them. Their intention was to preserve the relics of the dead by returning them to the dust, and to commemorate their life by a simple record of their name and age. The custom of making the graveyard a pleasure-ground is of modern origin. At the present time these old enclosures are shaded by a few trees that came up there without planting. The most common are the locust, the wild cherry, the velvet sumach, and the Lombardy poplar; and we have learned by habit to associate their rugged and homely appearance with the venerable objects that accompany them.

You can hardly conceive how much of the beauty of these ancient resting-places of the dead is due to the slate that forms the gravestones. Being of a dark color it harmonizes with nature; it is sober, but not sombre, and, unlike marble, it is often incrusted with lichens, and has no offensive glare. These are our only "rural cemeteries." Modern burying-grounds are but conservatories of sculpture and other works of decorative art. The use of white marble, be it ever so plain and simple, is incompatible with any idea of the picturesque. But when it is carved and embellished in the highest style of ornate art, we look upon the monuments as expressions of the vanity of the living under an ostentatious display of reverence for the dead.

As you continue your journey, the frequent changes in the course of the road are constantly varying your prospect. So little are these ways traversed that they are seldom defaced by repairs. The green rows of turf that mark their course have in many places seen fifty summers without disturbance. Now you are led a long distance in a straight direction over a plain, each side of the road being covered with whortleberry-bushes, loaded with fruit in its season, and you hear the halloos and frolic of children while employed in gathering it into baskets. On one of these levels you will often make half an hour's journey through a sparse growth of birches and pines, the ground being covered with wild-rose-bushes, crimson patches of lambkill, bayberry, sweet-fern, and blackberry-vines, the greensward glowing with the purple cranesbill, blue and white violets, and red summer lilies. This kind of scenery is always open and cheerful, for the sandy soil is dry and meagre, and supports but few large trees.

Where the road winds among the hills, the views it affords would charm any picturesque observer. It is seldom straight for more than a few hundred paces, and as you pass over the uneven grounds, you see the wood and shrubbery in every variety of grouping; for wild nature and the works of domestic art are mingled together more harmoniously in New England than in any other country. Sometimes the road separates into two parts, to meet again after leaving a long narrow ledge covered with wood, flowers, and ferns, and forming a perfect aviary of singing-birds. This is one of the objects that artistic improvement destroys, and then makes an absurd imitation of it in a city park or a private pleasure-ground; for if Fashion admires a scene in nature, she is still more delighted with its counterfeit. The road seldom passes over the top of the hill; it winds round it, unless it be a long ridge, when it is cut through it, the banks on each side being overhung by trees, with their roots half exposed from the sliding of the soil, the gravelly sides adorned with purple lupine, yellow St. John's wort, and the delicate flowers of the evening primrose, that open only at dewfall.

The road may soon carry you into the deep woods; and as the woods in New England, except those in bogs, stand chiefly upon the broken, hilly, and intractable parts of the surface, your course will be for a while through grounds as rugged as among the mountains and as picturesque as any mountain scenery in the world. It is delightful to emerge out of the darkness of these woods into an open valley containing a village of a few score houses, a church with a spire, a tavern, and a smithy, all enclosed by green and rugged hills. On a little grassy plain near the meeting of several roads leading from different points in the outskirts of the town stands the village school-house. It is a square building of one story, with a hurricane roof, painted red, and shaded by an elm. If it be summer, when the sons of the farmers are employed upon the land, and girls and small children only attend school, the teacher is a female, — a slender young woman, who has chosen the occupation of teaching, while her more buxom sisters are employed in active tasks at home.

The roads you have traversed are narrow and irregular, but all seem to terminate in this charming New England village, in which the simplicity of an earlier period is joined with the culture, refinement, and intelligence of the present day. Many enchanting scenes are assembled in it and hallow it; the plains are daisied with wild flowers, the surrounding hills are dressed in verdure and crowned with tall trees that seem like the guardians of its tranquillity. But the pride of the valley is this young teacher. The groves are but the arbor of which she is the sylph. Every circle in which she is imparadised is enlivened by her wit and beautified by her presence. Here you will remain and be happy, until ambition tempts you to join the tumult of commerce, and causes you to forget those sweet domestic scenes in which is enshrined all the happiness to be found in this world.

LIST OF ILLUSTRATIONS.

THE WOODS AND BY-WAYS OF NEW ENGLAND.

THE PRIMITIVE FOREST.

When the Pilgrim first landed on the coast of America, the most remarkable feature of its scenery that drew his attention, next to the absence of towns and villages, was an almost universal forest. A few openings were to be seen near the rivers, — immense peat-meadows covered with wild bushes and gramineous plants, interspersed with little wooded islets, and bordered on all sides by a rugged, silent, and dreary desert of woods. Partial clearings had likewise been made by the Indians for their rude hamlets, and some spaces had been opened by fire. But the greater part of the country was darkened by an umbrageous mass of trees and shrubbery, in whose gloomy shades were ever present dangers and bewilderment for the traveller. In these solitudes the axe of the woodman had never been heard, and the forest for thousands of years had been subject only to the spontaneous action of natural causes. To men who had been accustomed to the open and cultivated plains of Europe, this waste of woods, those hills without prospect, that pathless wilderness, and its inhabitants as savage as the aspect of the country, must have seemed equally sublime and terrible.

But when the colonists had cut roads through this desert, planted landmarks over the country, built houses upon its clearings, opened the hill-tops to a view of the surrounding prospect, and cheered the solitude by some gleams of civilization, then came the naturalist and the man of science to survey the aspect and productions of this new world. And when they made their first excursions over its rugged hills and through its wooded vales, we can easily imagine their transports at the sight of its peculiar scenery. How must the early botanist have exulted over this grand assemblage of plants, that bore resemblance to those of Europe only as the wild Indian resembles the fair-haired Saxon! Everywhere some rare herb put forth flowers at his feet, and trees of magnificent height and slender proportions intercepted his progress by their crowded numbers. The wood was so generally uninterrupted, that it was difficult to find a summit from which he could obtain a lookout of any considerable extent; but occasional natural openings exposed floral scenes that must have seemed like the work of enchantment. In the wet meadows were deep beds of moss of the finest verdure, which had seldom been disturbed by man or brute. On the uplands were vast fields of the checkerberry plant, social, like the European heath, and loaded half the year with its spicy scarlet fruit. Every valley presented some unknown vegetation to his sight, and every tangled path led him into a new scene of beauties and wonders. It must have seemed to him, when traversing this strange wilderness, that he had entered upon a new earth, in which nature had imitated, without repeating, the productions of his native East.

Along the level parts of New England and the adjacent country, wherever the rivers were languid in their course, and partially inundated their banks in the spring, were frequent natural meadows, not covered by trees, — the homes of the robin and the bobolink before the white man had opened to them new fields for their subsistence. In the borders of these openings, the woods in early summer were filled with a sweet and novel minstrelsy, contrasting delightfully with the silence of the deeper forest. The notes of the birds were wild variations of those which were familiar to the Pilgrim in his native land, and inspired him with delight amidst the all-prevailing sadness of woods that presented on the one hand scenes both grand and beautiful, and teemed on the other with horrors which only the pioneer of the desert could describe.

The whole continent, at the time of its discovery, from the coast to the Great American Desert, was one vast hunting-ground, where the nomadic inhabitants obtained their subsistence from the chase of countless herds of deer and buffalo. At this period the climate had not been modified by the operations of man upon the forest. It was less variable than now, and the temperature corresponded more definitely with the degrees of latitude. The winter was a season of more invariable cold, less interrupted by thaws. In New England and the other Northern States, snow fell in the early part of December, and lay on the ground until April, when the spring opened suddenly, and was not followed by those vicissitudes that mark the season at the present era. Such was the true forest climate. May-day came garlanded with flowers, lighted with sunshine, and breathing the odors of a true spring. It was then easy to foretell what the next season would be from its character the preceding years. Autumn was not then, as we have often seen it, extended into winter. The limits of each season were more precisely defined. The continent was annually visited by the Indian summer, that came, without fail, immediately after the fall of the leaf and the first hard frosts of November. This short season of mild and serene weather, the halcyon period of autumn, has disappeared with the primitive forest.

The original circumstances of the country have been entirely revolutionized. The American climate is now in that transition state which has been caused by opening the space to the winds from all quarters by operations which have not yet been carried to their extreme limit. These changes of the surface have probably increased the mean annual temperature of the whole country by permitting the direct rays of the sun to act upon a wider area, while they have multiplied those eccentricities of climate that balk our weather calculations at all seasons. There are still in many parts of the country large tracts of wood which have not been greatly disturbed. From the observation of these, and from descriptions by different writers of the last century, we may form a pretty fair estimate of the character and aspect of the forest before it was invaded by civilized man.

During this primitive condition of the country, the forest, having been left for centuries entirely to nature, would have formed a very intelligible geological chart. If we could have taken an extensive view of the New England forest, before any considerable inroads had been made by the early settlers, from an elevated stand on the coast, we should have beheld a dense and almost universal covering of trees. From this stand we might also trace the geological character of the soil, and its different degrees of fertility, dryness, and moisture, by the predominance of certain species and the absence of others. The undulations upon this vast ocean of foliage would come from the elevations and depressions of the ground; for the varying heights of the different assemblages of species upon the same level could hardly be perceived by a distant view. The lowest parts of this wooded region were at that period covered very generally with a crowded growth of the northern cypress, or white cedar. These evergreen swamps would constitute the darkest ground of the picture. The deep alluvial tracts would be known by the deciduous character of their woods and their lighter and brighter verdure, and the dry, sandy and diluvial plains and the gravelly hills and eminences by their white birches and tremulous poplars, their stunted pitch-pines and dwarfish junipers. For a century past the woods have been cleared mostly from the alluvial tracts; and the oaks, the hickories, the chestnuts, and other hard-wood trees, the primitive occupants of the rich and deep soils, have been succeeded in great measure by trees of softer wood, that originally grew on inferior land. The wooded aspect of the country cannot any longer be considered, as formerly, a good geological chart, except in some parts of Maine and the adjoining British Provinces.

One of the conditions most remarkable in a primitive forest is the universal dampness of the ground. The second growth of timber, especially if the surface were entirely cleared, stands upon a drier foundation. This greater dryness is caused by the absence of those vast accumulations of vegetable *débris* that rested on the ground before it was disturbed. A greater evaporation also takes place under the second growth, because the trees are of inferior size and stand more widely apart. Another character of a primitive forest is the crowded assemblage of trees and their undergrowth, causing great difficulty in traversing it. Innumerable straggling vines, many of them covered with thorns, like the green-brier, intercept our way. Immense trunks of trees, prostrated by hurricanes, lie in our path, and beds of moss of extreme thickness cover a great part of the surface, saturated with moisture. The trees are also covered with mosses, generated by the shade and dampness; and woody vines, like the climbing fern, the poison ivy, and the ampelopsis, fastened upon their trunks and trailing from their branches, make the wood in many places like the interior of a grotto. Above all, the traveller would notice the absence of those pleasant wood-paths that intersect all our familiar woods, and would find his way only by observing those natural appearances that serve as a compass to the Indian and the forester.

In primitive woods there is but a small proportion of perfectly formed trees; and these occur only in such places as permit some individuals to stand in an isolated position, and spread out their arms to their full capacity. When rambling in a wood we take note of several conditions which are favorable to this full expansion of their forms. On the borders of a lake, a prairie, or an open moor, or of an extensive quarry that projects above the soil, the trees will extend their branches into the opening; but as they are crowded on their inner side, they are only half developed. This expansion, however, is on the side that is exposed to view; hence the incomparable beauty of a wood on the borders of a lake or pond, on the banks of a river as viewed from the water, and on the circumference of a densely wooded islet.

Fissures and cavities are frequent in large rocks not covered with soil, allowing solitary trees which have taken root in them to acquire their full proportions. In such places, and on eminences that rise suddenly above the forest level, with precipitous sides, overtopping the surrounding woods, we find individual trees possessing the character of standards, like those we see by roadsides and in open fields. But perfectly formed trees can only be produced in openings and on isolated elevations such as I have described; and it is evident that these favorable circumstances must be rare. The trees in a forest are like those human beings who from their infancy have been confined in the workshops of a crowded manufacturing town, and who become closely assimilated and lose those marks of individual character by which they would be distinguished if they had been reared in a state of freedom and in the open country.

The primitive forest, in spite of its dampness, has always been subject to fires in dry seasons, which have sometimes extended over immense tracts of country. These fires were the dread of the early settlers, and countless lives have been destroyed by their flames often overwhelming entire villages. At the present time the causes of fire in the woods are very numerous; but before they were exposed to artificial sources of ignition it may have arisen from spontaneous combustion, caused by large accumulations of fermenting substances, or from lightning, or from the accidental friction of the trunks of half-prostrated trees crossing each other, and moved by a high wind. The forests in every part of the world have been subject to conflagrations; and there seems to be no other means that could be used by nature for removing old and worn-out forests, which contain more combustible materials than any young woods. The burned tracts in America are called barrens by the inhabitants; and as the vegetation on the surface is often entirely destroyed, the spontaneous renewal of it would display the gradual method of nature in restoring the forest. The successions of plants, from the beautiful crimson fireweed, through all the gradations of tender herbs, prickly bushes, and brambles, to shrubs and trees of inferior stature, until all, if the soil be deep and fertile, are supplanted by oaks, chestnuts, hickories, and other hardwood trees, are as regular and determinable as the courses of the planets or the orders of the seasons.

THE ASH.

It is interesting to note the changes that take place from one season to another in the comparative beauty of certain trees. The Ash, for example, during the early part of October, is one of the most beautiful trees of the forest, exceeded only by the maple in variety of tinting. In summer, too, but few trees surpass it in quality of foliage, disposed in flowing irregular masses, light and airy, but not thin, though allowing the branches to be traced through it, even to their extremities. It has a well-rounded head, neither so regular as to be formal, nor so broken as to detract from its peculiar grace. When standing with other trees in midsummer, in the border of a wood, or mingled with the standards by the roadside, the Ash would be sure to attract admiration. But no sooner have the leaves fallen from its branches than it takes rank below almost all other trees, presenting a stiff, blunt, and awkward spray, and an entire want of that elegance it affects at other seasons.

The Ash is a favorite in Europe, though deficient there in autumnal tints. It is a tree of the first magnitude, and has been styled in classical poetry the Venus of the forest, from the general beauty of its proportions and flowing robes. The English, however, complain of the Ash, on account of its tardy leafing in the spring and its premature denudation in the autumn. "Its leaf," says Gilpin, "is much tenderer than that of the oak, and sooner receives impression from the winds and frost. Instead of contributing its tint, therefore, in the wane of

the year, among the many colored offspring of the woods, it shrinks from the blast, drops its leaf, and in each scene where it predominates leaves wide blanks of desolate boughs amid foliage yet fresh and verdant. Before its decay we sometimes see its leaf tinged with a fine yellow, well contrasted with the neighboring greens. But this is one of nature's casual beauties. Much oftener its leaf decays in a dark, muddy, unpleasing tint."

The Ash is remarkable for a certain trimness and regularity of proportion, and it seldom displays any of those breaks so conspicuous in the outlines of the hickory, which in many points it resembles. The trunk rises to more than an average height before it is subdivided; but we do not see the central shaft above this subdivision, as in the poplar and the fir. Lateral branches seldom shoot from the trunk, save, as I have sometimes observed, a sort of bushy growth, surrounding it a little below the angles made by the lower branches. It is called in Europe "the painters' tree." But George Barnard, alluding to this fact, remarks: "Unlike the oak, the Ash does not increase in picturesqueness with old age. The foliage becomes rare and meagre, and its branches, instead of hanging loosely, often start away in disagreeable forms."

North America contains a greater number of species of the genus *Fraxinus* than any other part of the globe. But three of these only are common in New England,— the white, the red, and the black Ash. The first is the most frequent both in the forest and by the roadsides, the most beautiful, and the most valuable for its timber. All the species have pinnate and opposite leaves, and opposite branches in all the recent growth; but as the tree increases in size, one of the two invariably becomes abortive, so that we perceive this opposite character only in the spray. The leaflets are mostly in sevens, not so large nor so unequal as in the similar foliage of the hickory.

The white and the red Ash have so nearly the same external characters, that it requires some study to distinguish them. They do not differ in their ramification, nor in their autumnal hues. The black Ash may be readily identified by the leaves, which are sessile, and like those of the elder; also by the dark bluish color of the buds and newly formed branches, and the slenderness of its proportions. It seldom attains a great height or size, and is chiefly confined to swamps and muddy soils. The wood of this species is remarkable for strength and elasticity. The remarks of George Barnard respecting the localities of the Ash in Europe will apply to the American species: "Though seen everywhere, its favorite haunt is the mountain stream, where its branches hang gracefully over the water, adding much beauty to the scene. It is to be met with in every romantic glen and glade, now clinging with half-covered roots to a steep, overhanging cliff, and breaking with its light, elegant foliage the otherwise too abrupt line, or with its soft warm green relieving the monotonous coloring of the rocks or the sombre gray of some old ruin."

There are some remarkable superstitions and traditionary notions connected with the Ash-tree. The idea that it is offensive, and even fatal, to serpents, is not of modern origin, though not a rustic laborer can be found who would not consider an Ash-tree planted before his house as a charm against their intrusion. According to Pliny, if a serpent be surrounded on one side by fire and on the other by a barricade of the leaves and branches of the Ash-tree, he will escape through the fire, rather than through its fatal boughs. It is related in the Edda that man was first created from the wood of this tree, and it is not improbable that this superstition has some connection with the fable of Adam and Eve, and through this with the supposed antipathy of the serpent for the Ash-tree.

There is a saying in Great Britain, that, if the Ash puts forth its leaves before the oak, the following summer will be wet; but if the leafing of the oak precedes that of the Ash, it will be dry. I am not aware that any such maxim has obtained credence in the United States.

ANIMALS OF THE PRIMITIVE FOREST.

EUROPEAN travellers in this country frequently allude to the American forest as remarkable for its solitude and deficiency of animal life. Captain Hardy remarks that a foreigner is struck with surprise, when rambling through the bush, at the scarcity of birds, rabbits, and hares, and is astonished when in the deepest recesses of the wild country he sees but little increase of their numbers. When paddling his canoe through lake and river, he will startle but few pairs of exceedingly timid waterfowl where in Europe they swarm in multitudes. This scarcity of animals, I would remark, is not peculiar to the American wilderness. The same fact has been observed in extensive forests both in Europe and Asia; and in proportion as the traveller penetrates into their interiors he finds a smaller number of animals of almost every species. Birds, insects, and quadrupeds will multiply, like human beings, in a certain ratio with the progress of agriculture, so long as there remains a sufficiency of wild wood to afford them a refuge and a home. They use the forest chiefly for shelter, and the open grounds for forage; the woods are their house, the meadows their farm.

I had an opportunity for observing these facts very early in life, when making a pedestrian tour through several of the States. I commenced my journey in autumn, and being alone, I was led to take note of many things which, had any one accompanied me, would have escaped my observation. After passing a few weeks of the winter in Nashville, I directed my course through Tennessee and Virginia, and was often led through extensive ranges of forest. I never saw birds in any part of the United States so numerous as in the woods adjoining the city of Nashville, which was surrounded with immense corn-fields and cotton plantations. But while walking through the country I could not help observing the scarcity of birds and small quadrupeds in the woods whenever I was at a long distance from any village or habitation. Sometimes night would draw near before I had reached a hamlet or farm-house, where I might take lodging. On such occasions the silence of the woods increased my anxiety, which was immediately relieved upon hearing the cardinal or the mocking-bird, whose cheerful notes always indicated my approach to cultivated fields and farms.

That this scarcity of animal life is not peculiar to the American forest we have the testimony of St. Pierre, who says of the singing birds: "It is very remarkable that all over the globe they discover an instinct which attracts them to the habitations of man. If there be but a single hut in the forest, all the singing birds of the vicinity come and settle round it. Nay, they are not to be found except in places which are inhabited. I have travelled more than six hundred leagues through the forests of Russia, but never met with small birds except in the neighborhood of villages. On making the tour of fortified places in Russian Finland with the general officers of the corps of engineers with which I served, we travelled sometimes at the rate of twenty leagues a day without seeing on the road either village or bird. But when we perceived the sparrows fluttering about, we concluded we must be near some inhabited place. In this indication we were never once deceived."

It may be remarked, however, that birds and quadrupeds do not seek the company of man when they congregate near his habitations. They are attracted by the increased amount of all their means of subsistence that follows the cultivation of the land. The granivorous birds, no less than the insect-feeders, are benefited by the extension of agriculture. Even if no cereal grains were raised, the cultivated fields would supply them, in the product of weeds alone, more sustenance than a hundred times the same area in forest. Before there were any settlements of white men in this country, birds and small quadrupeds must have congregated chiefly about the wooded borders of prairies, on the banks of rivers, in fens and cranberry meadows, and around the villages of the red man. Their numbers over the whole continent were probably much smaller than at the present time, notwithstanding the merciless destruction of them by gunners and trappers.

There are but few tribes of animals that may be supposed to thrive only in the wild forest; and even these, if unmolested by man, would always find a better subsistence in a half-cultivated country abounding in woods of sufficient extent to afford them shelter and a nursery for their young, than in a continuous wilderness. Beasts of prey, however, are destroyed by man in the vicinity of all his settlements, to protect himself and his property from their attacks, and game-birds and animals of the chase are recklessly hunted both for profit and amusement. In Europe the clearing of the original forest was so gradual that the wild animals multiplied more rapidly with the progress of agriculture. Civilization advanced so slowly, and the arts made such tardy and gradual progress, that all species enjoyed considerable immunity from man. The game-birds and animals of the chase were not only preserved in forests attached to princely estates, but they were also protected by game-laws at a time when such laws were less needful because so few of the peasantry were accustomed to the use of the gun.

While the royal forests yielded these creatures a shelter and abode, the cultivated lands near their bounds afforded them subsistence; and they must have multiplied more rapidly in proportion to the increase of human population than in America after its settlement, where very different circumstances and events were witnessed.

America was colonized and occupied by civilized people, and the forests were swept away with a rapidity unprecedented in the history of man. Every pioneer was a hunter provided with guns and ammunition; every male member of his family over seven years of age was a gunner and a trapper. The sparse inhabitants of the forest, which if unmolested, as in the early period of European civilization, would have multiplied in proportion to their increased means of subsistence, have been, on the contrary, shot by the gunner, insnared by the trapper, and wantonly destroyed by boys for amusement, until some species have been nearly exterminated. Instead of increasing in a ratio with the supplies of their natural food, many tribes of them are now more scarce than they were in the primitive forest. The small birds alone, whose prolific habits and diminutive size were their protection, have greatly multiplied.

But even if birds and quadrupeds were unmolested by man, there are some tribes that would prefer to reside in the deep wood, while others would fix their abode in orchards and gardens. The wild pigeon has not been favored in any respect by the clearing of the forest. The food of this species is abundantly supplied in the wilds of nature in the product of beechen woods, hazel copses, groves of the chinquapin oak, and of the shores of lakes and arms of the sea covered with Canada rice and the maritime pea-vine. Their immense powers of flight enable them to transport themselves to new feeding-grounds after any present stock is exhausted, and to wing their way over hundreds of miles between their different repasts. This cannot be said of the grouse, the turkey, and the partridge, whose feeble powers of flight confine them to a narrow extent of territory; and these birds must have been frequently robbed of their farinaceous stores by flocks of wild pigeons during their itinerant foraging.

There are many species of birds which we associate with the wild-wood because they breed and find shelter there, but if we watched their habits we should learn that even these solitary birds make the cultivated grounds their principal feeding-places. Such are the quail, the partridge, and very many of our game-birds. The quail and the partridge are omnivorous, but, like our common poultry, are more eager to seize a grub or an insect than a grain of corn. A potato-field is hardly less valuable to a flock of quails than a field of corn, and affords more sustenance to the snipe and the woodcock than any other grounds. But these birds, as well as others, have diminished as those natural advantages have increased that should promote their multiplication.

Even our sylvias and thrushes, the most timid of all the winged tribe, birds hardly ever seen except in lonely woods, multiply with the clearing of the country and the increased abundance of their insect food. The vesper thrushes, that shun the presence of man, and will become silent in their musical evening if the rustling of the bushes indicates the approach of a human footstep, are more numerous in the woods of Cambridge than in any other part of the country. These are chiefly of maple, filled with underbrush, and afford the birds a harbor and a shelter, while the adjoining fields, in a state of the highest tillage, supply them plentifully with their natural food, consisting of worms and the larvæ of insects. The timid habits of these solitary birds are their chief protection. They will not expose themselves to observation, and on the approach of a human being they flee to the woods, where they are entirely concealed from the youths who destroy all sorts of small game. Birds of this species continue to grow more numerous, while the red thrush and the catbird are constantly diminishing in numbers because they breed outside of the wood, where they are more easily discovered.

American hares multiply as the forest is cleared, in spite of the unremitting persecution they suffer. The clover-fields adjoining the wood yield them a forage greatly superior to the scanty browsing of the shrubs and herbaceous plants of the forest. Deer would be favored by the same conditions, if they were not driven by the hunter into the most savage regions. Attempts are constantly made by our different State governments to protect these valuable birds and animals; but their acts must always be unavailing. The only means that will save them from extermination must be afforded by the universal establishment of forest conservatories, set apart exclusively for their protection.

THE AZALEA, OR SWAMP HONEYSUCKLE.

THE Azaleas are favorite flowering shrubs in florists' collections at the present day, and are remarkable for the delicacy of their flowers and the purity of their colors. In New England are only two species, — the Swamp Honeysuckle and the colored Azalea, a prostrate shrub bearing pink flowers. It cannot be doubted that the interest attached to a flower is greatly increased by finding it in the wild-wood. I have frequently observed this effect and the opposite upon suddenly meeting a garden flower in a field or wood-path, or a wild flower in the garden. When the Swamp Honeysuckle is seen growing with the fairer Azaleas of the florists in cultivated grounds, its inferiority is most painfully apparent; but when I encounter it in some green solitary dell in the forest, bending over the still waters, where all the scenes remind me only of nature, I am affected with more pleasure than by a display of the more beautiful species in a garden or greenhouse.

The Swamp Honeysuckle is one of the most interesting of the New England flowering shrubs, and a very well known species. It comes into flower about the first of July, and is recognized by its fragrance, — resembling that of the marvel of Peru, — by the similarity of its flowers to those of the woodbine, and their glutinous surface. It is found only in wet places, and delights in suspending its flowers over a gently flowing stream, the brink of a pool, or the margin of a pond, blending its odors with those of water-lilies, and borrowing a charm from the reflection of its own beauty on the surface of the still water. Though it bears no fruit, every rambler in the woods is grateful for the perfume it sheds around him while wandering in quest of its flowers. These are extremely delicate in texture and closely resemble those of the common white honeysuckle or woodbine of our gardens, not only in their general shape, but also in the appearance of several wilted flowers in the same cluster with perfect flowers and buds. A pulpy excrescence is often attached to this plant, which is familiarly known by the name of "swamp apple." It is slightly acidulous and sweet, and, though nearly insipid, is not disagreeable in flavor.

A more beautiful but less common species, with pale crimson flowers, is found in certain localities, that tends to multiply into varieties. It is a smaller shrub than the white Azalea, and does not show the same preference for wet places. All the species are more remarkable for their flowers than their foliage, which is of a pale glaucous green and small in quantity.

THE CANADIAN RHODORA.

In the latter part of May, when the early spring flowers are just beginning to fade, and when the leaves of the forest trees are sufficiently expanded to display all the tints attending the infancy of their growth, no plant attracts more admiration than the Canadian Rhodora. The flowers, of a purple crimson, are in umbels on the ends of the branches, appearing before the leaves. The corolla, consisting of long narrow petals, very deeply cleft, the stamens on slender hairy filaments, and the projecting style, resemble tufts of colored silken fringe. The Rhodora is from two to six feet in height, and is one of the most conspicuous ornaments of wet, bushy pastures in this part of the country. It is the last in the train of the delicate flowers of spring, and by its glowing hues indicates the coming of a brighter vegetation. When other shrubs of different species are only half covered with foliage, the Rhodora spreads out its flowers upon the surface of the variegated ground, in plats and clumps of irregular sizes, and sheds a checkered glow of crimson over whole acres of moor. The poets have said but little of this flower because it wants individuality. We look upon the blossoms of the Rhodora as we look upon the crimsoned clouds, admiring their general glow, not the cast of single flowers. But there is something very poetical in the rosy wreaths it affixes to the brows of Nature, still pallid with the long confinement of winter.

THE PASTORAL AND ROMANTIC.

It is usual to refer the sensations produced by the different objects of nature to some one of the general heads of the sublime, the beautiful, or the picturesque. All these terms are exceedingly vague, expressing, without clearly distinguishing, a great diversity of feelings and sentiments. By separating the ideas conveyed by them into more specific divisions, though we do not thereby escape a certain vagueness of signification that attaches to all metaphysical terms, we render them more distinct and intelligible. The word "picturesque" will not express the character of all those objects which could not be correctly described either as beautiful or sublime. There are descriptions of scenery that may properly be denominated pastoral and romantic, others rude, dreary, and desolate. Romantic scenery is usually described as that which is naturally fitted for adventure. Such is all abrupt and mountainous country, interspersed with woods and ravines favorable to escapes from danger and adapted to concealment.

In a painting or romance the most interesting person is the one who innocently suffers the greatest misfortunes, and individuals of high station and seeming prosperity can become objects of romantic interest only by exposure to some threatening danger, or by actual misfortune. A painting that should represent a lady in her parlor, surrounded by the luxuries and refinements of fashionable life, might elicit admiration, but could awaken no poetic interest. We may admire her beauty, the splendor that surrounds her, and the skill of the painter who devised the scene. But there is no poetry in the simple feeling of admiration, and what we merely admire seldom affects the heart. A mother sitting upon a solitary shore, with a group of young children clinging about her, looking for an approaching sail, is a scene that excites no admiration, but keenly awakens our sympathies, and is poetical in the highest degree. From humble life, or from greatness reduced to misfortune, the painter and the poet must select all those images and incidents that will deeply affect the soul.

In an old edition of the "Lady of the Lake," a poem full of romantic scenes, the frontispiece represents the heroine of the story alone in a skiff, near the shore of the lake. The royal hunter, having been separated from his companions, and being in a wild and lonely situation, by the side of Loch Katrine, sounds his bugle. This alarms the maiden, who quickly, on perceiving the hunter, pushes her light shallop from the shore. Ellen was a chieftain's daughter, and being alone in a skiff, near the margin of a solitary lake in the forest, she becomes an object of intensely romantic interest; and her youth and her beauty, her loneliness and her danger, yield a deeply picturesque and poetical character to the scene. Neither her beauty nor her rank would so deeply affect us, if there were nothing in her situation to awaken our sympathy and arouse our apprehension of her dangers. But when she is seen with hasty oar in the act of pushing her shallop into deeper water, to avoid the stranger huntsman, she becomes a romantic object in proportion to her beauty and her perils.

It is the affecting character of written or painted scenes of humble life that awakens the interest which has always been felt in the narrative of the "Deserted Village," — that perfect example of the pastoral elegy. Whether it be that we can more easily sympathize with the poor and humble, or that we feel that there is more happiness in a rustic cottage, where content resides, than in a mansion where constant endeavors are made to maintain a false appearance of superiority, it will not be denied that there is a charm in the pictures of humble life that cannot be transferred to those of the mansion or the palace. I cannot perceive that envy enters into our feelings on the one hand, or benevolence on the other. The happy tranquillity that seems to reside in the cottage, and the disquiet which would surely attend us in the opposite condition, make all the difference.

But we must add to a humble picture something of a pastoral character to yield it still more effect. Here is a neat but rude cottage, with two or three smiling and ruddy children at the door. It seems to be the dwelling, not of beggars, but of simple laborers. A woodman is standing near, in the act of cutting a branch from a tree. Hard by are a few cows, some reposing near a shed, others quietly feeding on an adjoining slope. This scene derives interest chiefly from its pastoral suggestions. It is rendered picturesque by calling up images of rural peace and the happy life of a humble tiller of the soil. It possesses moral beauty in a high degree, because it wears a quiet look of contentment, and is associated with all those charming fancies so generally inspired by scenes of rustic simplicity.

We proceed on our ramble until we reach a little nook or recess in the wood. A high wall of granite, covered with ferns, bounds it on two sides; a pleasant wood stands in the rear, while in front it overlooks the region below. In the centre of this recess is a hermitage, occupied by a venerable recluse. The edifice and its inhabitant, viewed in connection with the native beauties of the place, form a genuine scene of the romantic cast, heightened by our ideas of the piety and devotion of the hermit; but it does not awaken our sympathies or interest our affections like a simple pastoral scene. In all these cases it is the imagination that lends the scene its charms. To a man of cold heart and inactive mind, nothing is picturesque, nothing is poetical, solemn, pastoral, or romantic. Hence it is not difficult to understand why a man of culture must have sources of happiness which are entirely hidden from the boor and the sensualist.

It is this rural sentiment, this love of shepherd life and its accompaniments, that causes our interest in pastoral poetry, which has little else in general to recommend it. Our love of shepherd life, which is almost purely ideal and practically delusive, is closely allied with our love of nature and simplicity. We associate the employments of a shepherd with all the pleasant imagery of freedom and leisure, with shady groves and the sylvan muse, with the rustic pipe and all the various scenes described in the idyl and the eclogue. A shepherd's life is probably very tiresome to the rural swain who is obliged to follow it, but no less interesting to the spectator, whose imagination calls up in connection with it gentle flocks, murmuring bees, flowery meads, and the sweet Menalian strains of the classical eclogue. But the ideas of happiness which are associated with rural life in general are far from delusive; and its occupations are usually attended with more personal freedom than the pursuits either of trade or ambition.

Though I have joined together, in this essay, the pastoral and the romantic, they are in many respects opposite sentiments,— the one relating to action and adventure, the other to quiet and seclusion. The pastoral has by custom acquired a more extended meaning than its original expression of shepherd life, and includes the feeling with which we regard almost all quiet occupations in the country. The association of the woods and fields with the rural deities of the ancients is a part of our modern sentiment of the pastoral and romantic. Keats has founded his poem of "Endymion" on this sentiment, which greatly magnifies the agreeable impressions we derive from natural objects. In the mind of Keats all nature was a paradise of rural deities and beautiful objects of human love.

We look upon nature with more depth of affection when we have learned to people all the groves, hills, and fountains with their appropriate deities. The stream that winds through the valley is the more beautiful when it proceeds as it were from the urn of the naiad; and the sounds that reverberate from the hills produce a more animated sensation when we listen to them as the voice of the solitary Echo banished to her shell. A constant use of mythological figures in our descriptions of nature would be tiresome and commonplace; but the frugal and ingenious coloring of those descriptions with mythological imagery is always agreeable and poetical.

THE WILLOW.

THE Willow is of all trees the most celebrated in romance and romantic history. Its habit of growing by the sides of lakes and rivers, and of spreading its long branches over wells in solitary pastures, has given it a peculiar significance in poetry as the accompaniment of pastoral scenes, and renders it one of the most interesting objects in landscape. Hence there is hardly a song of nature, a rustic lay of shepherds, a Latin eclogue, or any descriptive poem, that does not make frequent mention of the Willow. The piping sounds from wet places in the spring of the year, the songs of the earliest birds, and the hum of bees when they first go abroad after their winter's rest, are all delightfully associated with this tree. We breathe the perfume of its flowers before the meadows are spangled with violets, and when the crocus has just appeared in the gardens; and its early bloom makes it a conspicuous object when it comes forth under an April sky, gleaming with a drapery of golden verdure among the still naked trees of the forest and orchard.

When Spring has closed her delicate flowers, and the multitudes that crowd around the footsteps of May have yielded their places to the brighter host of June, the Willow scatters the golden aments that adorned it, and appears in the deeper garniture of its own green foliage. The hum of insects is no longer heard among the boughs in quest of honey, but the notes of the phebe and the summer yellow-bird, that love to nestle in their

spray, may be heard from their green shelter on all summer noons. The fresh and peculiar incense of the peat-meadows, with their purple beds of cranberry-vines and wild strawberries, the glistening of still waters, and the sight of little fishes that gambol in their clear depths, are circumstances that accompany the Willow, and magnify our pleasure on beholding it, either in a picture or real landscape. We prize the Willow for its material qualities no more than for its poetic relations; for it is not only the beauty of a tree, but the scenes with which it is allied, and the ideas and images it awakens in the mind, that make up its attractions.

The very name of this tree brings to mind at once a swarm of images, rural, poetical, and romantic. There is a softness in the sound of Willow that accords with the delicacy of its foliage and the flexibility of its slender branches. The syllables of this word must have been prompted by the mellow tones which are produced by the wind when gliding through its airy spray. Writers of romance have always assigned the Willow to youthful lovers, as affording the most appropriate arbor for their rustic vows, which would seem to acquire a peculiar sacredness when spoken under the shade of the most poetical of all trees.

The Willow, though tenacious of life, will not prosper in dry places. Its presence is a sure indication of water, either on the surface of the ground or a little beneath it. The grass is green at all times under this tree, and the herds that browse upon its foliage and young branches find beneath them the most grateful pasture. In the New England States it has long been customary to plant Willows by the wayside, wherever the road passes over wet grounds. Some of the most delightful retreats of the pedestrian are found under their shady boughs. When he is panting with heat and thirst, the sight of their green rows fills him with new animation, as they indicate the presence of water as well as cooling shade. The same comely rows are seen skirting the pools and watercourses of our pastoral hills and arable meadows. They are planted also by the sides of streams and canals, where they serve, by their long and numerous roots, to consolidate the banks, and by their leaves and branches afford shelter to cattle. These Willows are among the fairest ornaments of the landscape in Massachusetts just after the elm and red maple have put forth their flowers. And so lively is their appearance, with their light green foliage, that when we meet with a group of them in the turn of a road on a cloudy day, we seem to be greeted with a sudden gleam of sunshine.

The Willow is one of the few trees which have been transplanted from Europe to our own soil without being either equalled or surpassed by some American tree of kindred species. But there is no indigenous Willow in any part of the American continent that will bear comparison in size and in those general qualities which we admire in trees, either with the Weeping Willow or the common yellow Willow. The latter is as frequent in our land as any one of our native trees, except in the forest. It attains a considerable height and great dimensions, seldom forming a single trunk, but sending upward from the ground, or from a very short bole, three or four diverging branches, so as to resemble an immense shrub. This mode of growth is caused perhaps by our way of planting it,— by inserting into the ground cuttings which have no leading shoot. Indeed, all these Willows are pollards. Not one of the species is found in our forest, except where it has spread over land that has once been cleared and cultivated. In that case, we find mixed with the forest trees Willows, apple-trees, and lilacs, which were planted there before the tract was restored to nature. I have seen trees of this species growing as standards of immense size, with their branches always joining the trunk very near the ground. On this account little rustic seats and arbors are more frequently erected in the crotch of a Willow than in that of any other tree.

The most of our indigenous Willows are mere shrubs. Though there are above thirty American species, but few of them rise to the stature of trees. Some of them are creeping plants and prostrate shrubs, some are neat and elegant trees in miniature. Their branches are also of many colors, some of a fine golden hue, spreading a sort of illumination over the swamps where they abound; some are red; others with foliage so dark as to have gained the name of Mourning Willow. Some, like our common bog Willow, are called white, from their downy or silken aments. One of the most beautiful of the small species is the golden osier, or Basket Willow. The yellow twigs of this shrub, coming up from the ground like grass without subdivisions, but densely from one common root, are very ornamental to low grounds. It would seem as if Nature, who has given but little variety to the foliage of this tree, had made up for its deficiency by causing the different species to display a charming variety in their size. Thus, while the common yellow Willow equals the oak in magnitude, there are many species which are miniature shrubs, not larger than a heath plant. As one of the beautiful gifts of nature, the Willow claims a large share of our admiration. Though not a convenient ornament of our enclosures, the absence of this tree from the banks of quiet streams and glassy waterfalls, overhanging rivers and shading the brink of fountains, would be most painfully felt by every lover of nature.

ROTATION AND DISTRIBUTION.

It has been observed by foresters that there is a tendency in any soil which has long been occupied by a certain kind of timber, to produce, after the trees have been felled, a very different kind, if it be left to its spontaneous action. The laws affecting such rotations have been very well ascertained, and a careful investigation of the subject would undoubtedly reveal many curious facts not yet known. If the stumps of the trees, consisting of oak, ash, maple, and some other deciduous kinds, remain after the wood is felled, they will throw up suckers, and the succeeding timber will be an inferior growth of the original wood. But if the stumps and roots of the trees should be entirely removed, it would be more difficult to determine what would be the character of the next spontaneous growth. It would probably be planted by the kinds that prevail in the neighboring forests, and it would depend on the character of the soil whether the hard or soft wood trees would finally predominate.

There is an important chemical agency at work, that originally determines the distribution of forests, and afterwards their rotation. The hard-wood trees require more potash and a deeper soil than the coniferous and soft-wood trees. Hence they are found chiefly on alluvial plains and the lower slopes of mountains, where the soil is deep and abounds in all valuable ingredients for the support of vegetation. Pines and firs, on the contrary, though frequently discovered of an immense size on alluvial soils, are generally crowded out of such grounds by the superior vigor of the hard-wood trees; and they can only maintain their supremacy on barren and sandy levels, and the thin soils of mountain declivities, too meagre to support the growth of timber of superior kinds. But a wood must stand a great many years, several centuries perhaps, after its spontaneous restoration, before this order of nature could be fully established. We must observe the spontaneous growth and distribution of herbaceous plants in different soils to ascertain these laws, which are the same in a field as in a forest.

When any growth of hard wood has been felled and the whole removed from the ground, the soil, having been exhausted of its potash, cannot support a new and vigorous growth of the same kind of timber. The succession will consist of a meagre growth of the same species from seeds already planted there; but the white birch and poplar, especially the large American aspen, usually predominate in clearings in this part of the country. When a pine wood is felled, it is succeeded by an inferior growth of conifers, and a species of dwarf or scrub oak. Seldom, indeed, after any kind of wood has been cut down and carried away from the spot, can the exhausted soil support another that is not inferior in quality or species. Though an oak wood may be succeeded by pines, a pine wood will not be succeeded by oaks or any other hard timber, unless the trees were burned and their ashes restored to the soil. Hence we may account for the fact that poplars, white birches, and wild-cherry-trees, occupy a larger proportion of the ground that is now covered with wood than they did a century ago, in all parts of the country.

I have already alluded to the well-known fact, that the generic character of the timber, in the distribution of the primitive forest, in any country, is determined in great measure by the geological character of the soil. On sandy plains in the primitive forest, the white birch, the poplar, the aspen, and the pitch pine were abundant, as they are now on similar soils. The preference of the red maple for wet and miry soils is well known; while hard maple, oak, beech, and hickory do not prosper except in strong alluvial tracts. A heavy growth of hard timber indicates a superior soil; pine indicates an inferior one, if it has been left to the spontaneous action of nature. In the primitive forest we were sure of finding such relations of soil and species. They are not so invariable since the operations of agriculture have interrupted the true method of nature.

When a wood has been burned, the process of renewal, when left to nature, is much more tardy than if it had been felled, since it can now be restored only by a regular series of vegetable species, which must precede it, according to certain inevitable laws. The soil, however, being improved and fertilized by the ashes of the burnt timber, is in a chemical condition to support a luxuriant forest as soon as in the course of nature it can be planted there. Trees will not immediately come up from this burnt ground as in a clearing; and if they should appear, they would mostly perish from the want of protection. In the order of nature herbaceous plants are the first to occupy the soil, and these are followed by a uniform succession of different species. There is an epilobium, or willow herb, with elegant spikes of purple flowers, conspicuous in our meadows in August, which is one of the earliest occupants of burnt ground, hence called fireweed in Maine and Nova Scotia. The downy appendage to its seeds causes it to be planted there by the winds immediately after the burning. The trillium appears also in great abundance upon the blackened surface of the ground in all wet places. Plants like the ginseng, the erythronium, and the like, whose bulbs or tubers lie buried deep in the mould, escape destruction, and come up anew. These, along with several compound plants with downy seeds, and a few ferns and equisetums, are the first occupants of burnt lands.

But the plants mentioned above have no tendency to foster the growth of young trees. They are, however, succeeded by the thistles and thorny plants, which are nature's preparation of any tract, once entirely stripped of vegetation, as a nursery for the seedlings. All the phenomena of nature's rotation are but the necessary giving place of rapid-growing and short-lived plants to others which are perennial and more capable of maintaining their ground after being once planted. Thorns and thistles soon appear on burnt lands, and protect the young trees as they spring up, both from the winds and the browsing of animals. Thus many an oak has been nursed in a cradle of thorns and brambles, and many a lime-tree growing in a bower of eglantine has been protected by its thorns from the browsing of the goat.

We very early discover a variety of those woody plants that bear an edible fruit, which is eaten by birds and scattered by them over the land, including many species of bramble. The fruit-bearing shrubs always precede the fruit-bearing trees; but the burnt land is first occupied by those kinds that bear a stone-fruit. Hence great numbers of cherry-trees and wild-plum-trees are found there, as the natural successors of the wild gooseberry and bramble-bushes. These are soon mixed with poplars, limes, and other trees with volatile seeds. But oaks, hickories, and the nut-bearing trees must wait to be planted by squirrels and field-mice and some species of birds. The nut-bearers, therefore, will be the last to appear in a burnt region, for the little quadrupeds that feed upon their fruit will not frequent this spot until it is well covered with shrubbery and other vegetation. If the soil be adapted to the growth of heavy timber, the superior kinds, like the oak, the beech, and the hard maple, will gradually starve out the inferior species, and in the course of time predominate over the whole surface.

When I consider all these relations between plants and animals, I feel assured, if the latter were destroyed that plant their seeds, many species would perish and disappear from the face of the earth. Nature has provided, in all cases, against the destruction of plants, by endowing the animals that consume their fruits with certain habits that tend to perpetuate and preserve them. In this way they make amends for the vast quantities they consume. After the squirrels and jays have hoarded nuts for future use, they do not find all their stores; and they sow by these accidents more seeds than could have been planted by other accidental means, if no living creature fed upon them. Animals are not more dependent on the fruit of these trees for their subsistence, than the trees are upon them for the continuance of their species. And it is pleasant to note that, while plants depend on insects for the fertilization of their flowers, they are equally indebted to a higher order of animals for planting their seeds. The wasteful habits of animals are an important means for promoting this end. The fruit of the oak, the hickory, and the chestnut will soon decay if it lies on the surface of the ground, exposed to alternate dryness and moisture, and lose its power of germination. Only those nuts which are buried under the surface are in a condition to germinate. Many a hickory has grown from a nut deposited in the burrow of a squirrel; and it is not an extravagant supposition that whole forests of oaks and hickories may have been planted in this manner.

These facts are too much neglected in our studies of nature. A knowledge of them, and a consideration of their bearings in the economy of nature, might have saved many a once fertile country from being converted into a barren waste, and may serve yet to restore such regions to their former happy condition. But these little facts are not of sufficient magnitude to excite our admiration, and they involve a certain process of reasoning that is not agreeable to common minds, or even to the more cultivated, which have been confined chiefly to technology. The few facts to which I have alluded in this essay are such as lie at the vestibule of a vast temple that has not yet been entered. I am not ready to say that no single species of the animal creation may not be destroyed without derangement of the method of nature; for thousands have, in the course of time, become extinct by the spontaneous action of natural agents. But there is reason to believe that, if any species should be destroyed by artificial means, certain evils of grievous magnitude might follow their destruction.

The frugivorous birds are the victims of constant persecution from the proprietors of fruit gardens. Their persecutors do not consider that their feeding habits have preserved the trees and shrubs that bear fruit from utter annihilation. They are the agents of nature for distributing vegetables of all kinds that bear a pulpy fruit in places entirely inaccessible to their seeds by any other means. Notwithstanding the strong digestive organs of birds, which are capable of dissolving some of the hardest substances, the stony seeds of almost all kinds of pulpy fruit pass through them undigested. By this providence of nature the whole earth is planted with fruit-bearing trees, shrubs, and herbaceous plants, while without it these would ultimately become extinct. This may seem an unwarrantable assertion. It is admitted that birds alone could distribute the seeds of this kind of plants upon the tops of mountains and certain inaccessible declivities, which, without their agency, must be entirely destitute of this description of vegetation. But these inaccessible places are no more dependent on the birds than the plains and the valleys. The difference in the two cases is simply that the one is apparent, like a simple proposition in geometry, and the other requires a course of philosophical reasoning to be perfectly understood.

THE WEEPING WILLOW.

In the early part of my life, one of my favorite resorts during my rambles was a green lane bordered by a rude stone wall, leading through a vista of overarching trees, and redolent always with the peculiar odors of the season. At the termination of this rustic by-road, — a fit approach to the dwelling of the wood-nymphs, — there was a gentle rising ground, forming a small tract of table-land, on which a venerable Weeping Willow stood, — a solitary tree overlooking a growth of humble shrubs, once the tenants of an ancient garden. The sight of this tree always affected me with sadness mingled with a sensation of grandeur. This old solitary standard, with a few rose-bushes and lilacs beneath its umbrage, was all that remained on the premises of an old mansion-house which had long ago disappeared from its enclosure. Thus the Weeping Willow became associated in my memory, not with the graveyard or the pleasure-ground, but with these domestic ruins, the sites of old homesteads whose grounds had partially reverted to their primitive state of wildness.

Of all the drooping trees the Weeping Willow is the most remarkable, from the perfect pendulous character of its spray. It is also consecrated to the Muse by the part which has been assigned to it in many a scene of romance, and by its connection with pathetic incidents recorded in Holy Writ. It is invested with a moral interest by its symbolical representation of sorrow, in the drooping of its terminal spray, by its fanciful use as a garland for disappointed lovers, and by the employment of it in burial-grounds and in funereal paintings. We remember it in sacred history, associating it with the rivers of Babylon and with the tears of the children of Israel, who sat down under the shade of this tree and hung their harps upon its branches. It is distinguished by the graceful beauty of its outlines, its light green delicate foliagē, its sorrowing attitude, and its flowing drapery.

Hence the Weeping Willow never fails to please the sight even of the most insensible observer. Whether we see it waving its long branches over some pleasure-ground, overshadowing the gravel-walk and the flower garden, or watching over a tomb in the graveyard, where the warm hues of its foliage yield cheerfulness to the scenes of mourning, or trailing its floating branches, like the tresses of a Naiad, over some silvery lake or stream, it is in all cases a beautiful object, always poetical, always picturesque, and serves by its alliance with what is hallowed in romance to bind us more closely to nature.

It is not easy to imagine anything of this character more beautiful than the spray of the Weeping Willow. Indeed, there is no other tree that is comparable with it in this respect. The American elm displays a more graceful bend of all the branches that form its hemispherical head; and there are several weeping birches which are very picturesque when standing by a natural fountain on some green hillside. The river maple is also a theme of constant admiration, from the graceful flow of its long branches that droop perpendicularly when laden with foliage, but partly resume their erect position in winter, when denuded. But the style of all these trees differs entirely from that of the Weeping Willow, which in its peculiar form of beauty is unrivalled in the whole vegetable kingdom.

It is probable that the drooping trees acquired the name of "weeping," by assuming the attitude of a person in tears, who bends over and seems to droop. This is the general attitude of affliction in allegorical representations. But this habit is far from giving them a melancholy expression, which is more generally the effect of dark sombre foliage. Hence the yew seems to be a more appropriate tree for burial-grounds, if it be desirable to select one of a sombre appearance. The bending forms of vegetation are universally attractive, by emblemizing humility and other qualities that excite our sympathy. All the drooping plants, herbs, trees, and shrubs are poetical, if not picturesque. Thus lilies, with less positive beauty, are more interesting than tulips.

A peculiar type of the drooping tree is seen in the fir, whose lower branches bend downwards, almost without a curve, from their junction with the stem of the tree. This drooping is caused by the weight of the snow that rests upon the firs during the winter in their native northern regions. There is a variety of the beech, and another of the ash, which has received the appellation of *weeping*, from an entire inversion of the branches, both large and small. Such trees seem to me only a hideous monstrosity, and I never behold them without some disagreeable feelings, as when I look upon a deformed animal.

VERNAL WOOD-SCENERY.

ALL the seasons display some peculiar beauty that comes from the tints as well as the forms of vegetation. Even the different months have their distinguishing shades of light and color. Nature, after the repose of winter, very slowly unfolds her beauties, and is not lavish in the early months of any description of ornament. Day by day she discloses the verdure of the plain, the swelling buds with their lively and various colors, and the pale hues of the early flowers. She brings along her offerings one by one, leading from harmony to harmony, as early twilight ushers in the ruddy tints of morn. We perceive both on the earth and in the skies the forms and tints that signalize the revival of Nature, and every rosy-bosomed cloud gives promise of approaching gladness and beauty.

By the frequent changes that mark the aspect of the year we are preserved at all times in a condition to receive pleasure from the outward forms of Nature. Her tints are as various as the forms of her productions; and though spring and autumn, when the hues of vegetation are more widely spread and yield more character to the landscape, are the most remarkable for their general beauty, individual objects in summer are brighter and more beautiful than any that can be found at other times. In the early part of the year, Nature tips her productions with softer hues, that gradually ripen into darker shades of the same color, or into pure verdure. By pleasant and slow degrees she mingles with the greenness of the

plain the hues of the early flowers, and spreads a charming variety of warm and mellow tints upon the surface of the wood.

In treating of vernal tints, I shall refer chiefly to effects produced, without the agency of flowers, by that general coloring of the leaves and spray which may be considered the counterpart of the splendor of autumn. In the opening of the year many inconspicuous plants are brought suddenly into notice by their lively contrast with the dark and faded complexion of the ground. The mosses, lichens, and liverworts perform, therefore, an important part in the limning of the vernal landscape. On the bald hills the surfaces of rocks that project above the soil, and are covered with these plants, are brighter than the turf that surrounds them, with its seared grasses and herbage. They display circles of painted lichens, varying from an olive-gray to red and yellow, and tufts of green mosses which surpass the fairest artificial lawn in the perfection of their verdure. Many of the flowerless plants are evergreen, especially the ferns and lycopodiums, and nearly all are earlier than the higher forms of vegetation in ripening their peculiar hues.

The first remarkable vernal tinting of the forest is manifest in the spray of different trees. As soon as the sap begins to flow, every little twig becomes brightened on the surface, as if it had been glossed by art. The swelling of the bark occasioned by the flow of sap gives the whole mass a livelier hue. This appearance is very evident in the peach-tree, in willows and poplars, in the snowy mespilus, and in all trees with a long and slender spray. Hence the ashen green of the poplar, the golden green of the willow, and the dark crimson of the peach-tree, the wild rose, and the red osier, are perceptibly heightened by the first warm days of spring. Nor is this illumination confined to the species I have named; for even the dull sprays of the apple-tree, the cherry, the birch, and the lime, are dimly flushed with the hue of reviving life. As many of the forest trees display their principal beauty of form while in their denuded state, this seasonal polish invites our attention, particularly to those with long and graceful branches.

The swelling buds, which are for the most part very highly colored, whether they enclose a leaf or a flower, add greatly to this luminous appearance of the trees. These masses of innumerable buds, though mere colored dots, produce in the aggregate a great amount of color. This is apparent in all trees as soon as they are affected by the warmth of the season. But as vegetation comes forward, the flower-buds grow brighter and brighter, till they are fully expanded, some in the form of fringes, as in most of our forest trees, others, as in our orchard trees, in clusters of perfect flowers. This drapery of fringe, seldom highly colored, but containing a great variety of pale shades, that hangs from the oak, the birch, the willow, the alder, and the poplar, is sufficient to characterize the whole forest, and forms one of the most remarkable phenomena of vernal wood-scenery.

It is generally supposed that the beauties of tinted foliage are peculiar to autumn. I do not recollect any landscape painting in which the tints of spring are represented. All the paintings of colored leaves are sketches of autumnal scenes, or of the warm glow of sunlight. Yet there is hardly a tree or a shrub that does not display in its opening leaves a pale shade of the same tints that distinguish the species or the individual tree at the time of the fall of the leaf. The birch and the poplar imitate in their half-developed leaves the yellow tints of their autumnal dress, forming a yellow shade of green. The tender leaves of the maple and of the different oaks are all greenish purple of different shades.

On the other hand, the foliage of trees that do not change their color in the autumn displays only a diluted shade of green, in its half-unfolded state. This remark, however, is not universal in its application; for we see the lilac, that appears in autumn without any change, coming out in the spring with dark impurpled foliage.

Green cannot, therefore, be said to characterize a vernal landscape. It belongs more especially to summer. The prevailing color of the forest during the unfolding of the leaf, when viewed from an elevated stand, is a cinereous purple, mingled with an olive-green. The flowers of the elm, of a dark maroon, and the crimson flowers of the red maple, coming before their leaves, are an important element in the earliest hues of the wood. The red maple, especially, which is the principal timber of the swamps in all the southern parts of New England, yields a warm and ruddy glow to the woods in spring, hardly less to be admired than its own bright tints in October. Green hues, which become, day by day, more apparent in the foliage, do not predominate until summer has arrived and is fully established.

It is only in the spring that the different species of the forest can be identified by their colors at distances too great for observing their botanical characters. A red-maple wood is distinguished by the very tinge that pervades the spray, when the trees are so far off that we cannot see the forms of their branches and flowers, as if the ruddy hues of morning illuminated the whole mass. A grove of limes would be known by their dark-colored spray, approaching to blackness; an assemblage of white birches by that of a chocolate-color diverging from their clean white shafts. A beechen grove would manifest a light cinereous color throughout, mixed with a pale green as the foliage appears. If there were as many assemblages as there are species, we might at the time the buds are starting see in each some shade to distinguish it from all the others. The different complexions of the woods, as observed in their spray no less than in their foliage at a later period, would form a curious and not uninteresting study.

THE HORSE-CHESTNUT.

THE Horse-Chestnut I would compare with the locust on account of their difference, not their resemblance. Like the locust, it is remarkable for the beauty of its flowers, though even in this respect the trees are of an opposite character; the one bears them in upright pyramids, the other in pendent racemes. Those of the locust are half closed and modest in their colors of white and brown; those of the Horse-Chestnut are wide open and somewhat flaring, though of a delicate rose-color and white. While in blossom the tree is unsurpassed in its beautiful display of flowers, that "give it the appearance of an immense chandelier covered with innumerable girandoles."

After all, we can bestow very little praise upon the Horse-Chestnut, except for its flowers. The foliage of the tree displays neither lightness, nor elegance, nor brilliancy of verdure, nor autumnal tinting, nor any flowing beauty of outline. On the contrary, it is homely and heavy, though it affords a very deep shade. Indeed, when we view a Horse-Chestnut from a moderate distance, the arrangement of its leaves give it a very pleasing tufted appearance, unlike what we see in any other species. George Barnard says of it: "This cannot be called a picturesque tree, its shape being very formal; but the broad masses of foliage, although too defined and unbroken to be agreeable to the painter, are grand and majestic when seen in an avenue or in groups."

As a shade-tree, or a tree for avenues and pleasure-grounds, none would deny the merits of the Horse-Chestnut; but when denuded it is a miserable-looking object, with its terminal branches resembling drumsticks, its primness without grace, and its amplitude without grandeur. The birds seldom build their nests among its branches, which are too wide apart to afford them protection or accommodation; for this tree is absolutely without any spray. Its fruit, which is borne in great abundance, sustains neither bird nor quadruped, nor is it profitable for man. Hence it has always been regarded by poets and moralists as a symbol of extravagance and waste.

THE CATALPA.

THE Catalpa, though an American tree, is not indigenous in New England, nor farther north than Philadelphia. It is allied, in its botanical characters, to the bignonia, one of the most magnificent of the American flowering vines, which in Virginia and the Carolinas climbs the trunks of the loftiest trees, and, rising to a hundred feet or more, completely encompasses them with flowers of rare beauty and foliage of the finest green. The Catalpa requires notice here, because it is not uncommon in our gardens and pleasure-grounds, and it is becoming more and more general as a wayside tree. It is remarkable as a late bloomer, putting forth its large panicles of white flowers late in July, when those of other trees and shrubs have mostly faded, and covering the tree so thickly as almost to conceal its dense mass of foliage. The leaves are very large, but flowing, heart-shaped, and of a light and somewhat yellowish green. The Catalpa is not yet very common; but it is one of those rare productions which is never seen without being admired.

FORMS AND EXPRESSION OF TREES.

THE different forms of trees, and their endless variety of foliage and spray, have, from the earliest times, been favorite studies of the painter and the naturalist. Not only has each species certain distinguishing marks, but their specific characters are greatly modified in individual trees. The Psalmist compares a godly man to a tree that is planted by rivers of water, whose leaf shall not wither, — seeing in the stateliness and beauty of such a tree an emblem of the noble virtues of the human heart. Trees are distinguished by their grandeur or their elegance, by their primness or their grace, by the stiffness of their leaves and branches or by their waving and tremulous motions. Some stand forth as if in defiance of the wind and the tempest; others, with long drooping branches, find security in bending to the gale, like the slender herbs in the meadow.

Trees are generally classed as landscape ornaments, according to their general outlines. "Some trees ascend vertically," says St. Pierre, "and having arrived at a certain height, in an air perfectly unobstructed, fork off in various tiers, and send out their branches horizontally, like an apple-tree; or incline them towards the earth, like a fir; or hollow them in the form of a cup, like the sassafras; or round them into the shape of a mushroom, like the pine; or straighten them into a pyramid, like the poplar; or roll them as wool upon the distaff, like the cypress; or suffer them to float at the discretion of the winds, like the birch." These are the normal varieties in the shape of trees. Others may be termed accidental, like those of the tall and imperfectly developed trees, which have been cramped by growing in dense assemblages, and of the pollards that have issued from the stumps and roots of other trees.

Trees are generally wanting in that kind of beauty which we admire in a vase, or an elegant piece of furniture. They have more of those qualities we look for in a picture and in the ruder works of architecture. Nature is neither geometrical nor precise in her delineations. She betrays a design in all her works, but never casts two objects in the same mould. She does not paint by formulas, nor build by square and compass, nor plant by a line and dibble; she takes no note of formal arrangements, or of the "line of beauty," or of direct adaptation of means to ends. She shakes all things together, as in a dice-box, and as they fall out there they remain, growing crooked or straight, mean or magnificent, beautiful or ugly, but adapted by the infinite variety of their forms and dispositions to the wants and habits of all creatures.

The beauty of trees is something that exists chiefly in our imagination. We admire them for their evident adaptation to purposes of shade and shelter. Some of them we regard as symbols or images of a fine poetic sentiment. Such are the slender willows and poplars, that remind us of grace and refinement, becoming the emblems of some agreeable moral affection, or the embodiment of some striking metaphor. Thus Coleridge personifies the white birch as the "Lady of the Woods," and the oak by other poets is called the monarch, and the ash the Venus of the forest. The weeping willow beautiful on account of its graceful spray, becomes still more so when regarded as the emblem of sorrow. The oak, in like manner, is interesting as the symbol of strength and fortitude. A young fir-tree always reminds us of primness; hence the name of spruce, which is applied to many of the species, is a word used to express formality. The cedar of Lebanon would be viewed by all with a certain romantic interest, on account of the frequent mention of it in Holy Writ, as well as for its nobleness of dimensions and stature.

It is with certain interesting scenes in the romance of travel that we associate the palms of the tropics. They have acquired singular attractions by appearing frequently in scenes that represent the life and manners of the simple inhabitants of the equatorial regions. We see them in pictures bending their fan-like heads majestically over the humble hut of the Indian, supplying him at once with milk, bread, and fruit, and affording him the luxury of their shade. They emblemize the beneficence of nature, which, by means of their products, supplies the wants of man before he has learned the arts of civilized life.

Writers in general apply the term "picturesque" to trees which are devoid of symmetry and very irregular in their outlines, either crooked from age or from some natural eccentricity of growth. Thus the tupelo is so called, to distinguish it from round-headed and symmetrical or beautiful trees. This distinction is not very precise; but it is sanctioned by general use, and answers very well for common purposes of vague description. I shall use the words in a similar manner, not adhering to the distinction as philosophical. Indeed, it is impossible to find words that will clearly express a complex idea. Words are very much like tunes played on a jew's-harp; the notes intended to be given by the performer are accompanied by the louder ring of the key-note of the instrument, making it difficult to detect the notes of the tune, except in the hands of an extraordinary performer.

Nature has provided against the disagreeable effects that would result from the dismemberment of trees, by giving to those which are the most common a great irregularity of outline, admitting of disproportion without deformity. Symmetry in the forms of natural objects becomes wearisome by making too great a demand upon the attention required for observing the order and relations of the different parts. But if the objects in the landscape be irregular, both in their forms and their distribution, we make no effort to attend to the relations of parts to the whole, because no such harmony is indicated. Such a scene has the beauty of repose. The opposite effect is observed in works of architecture, in which irregularity puzzles the mind to discover the mutual relations of parts, and becomes disagreeable by disturbing our calculations and disappointing our curiosity. The charm of art is variety combined with uniformity; the charm of nature is variety without uniformity. Nature speaks to us in prose, art in verse.

Though we always admire a perfectly symmetrical oak or elm, because such perfection is rare, it will be admitted that the irregular forms of trees are more productive of agreeable impressions on the mind. The oak, one of the most interesting of all trees, is, in an important sense, absolutely ugly, especially when old age has increased its picturesque attractions. Indeed, if we could always reason correctly on the subjects of our consciousness, we should find that a very small part of that complex quality which we call beauty yields any organic pleasure to the sight. The charm of most of the objects in this category exists only in our imaginations. In trees and the general objects of the landscape we look neither for symmetry nor proportion; the absence of these qualities is, therefore, never disagreeable. It is the nonfulfilment of some expectation, or the apparently imperfect supply of some important want, that offends the sight, as when a conspicuous gap occurs in some finely proportioned work of art.

The fantastic shapes assumed under certain conditions by the elm, the tupelo, the swamp-oak, and less frequently by the beech and the hickory, constitute one of the principal attractions of a half-wooded landscape, and never affect us with any sense of deformity.

THE LILAC.

THE Lilac, though not one of our native trees, has become so generally naturalized in our fields and gardens as hardly to be distinguished from them except by its absence from the forest. It is common in all waste lands that were formerly the sites of ancient dwelling-houses, marking the spot where the garden was situated by its irregular clumps; for when neglected it does not assume the shape of a tree, but forms an assemblage of long stems from one spreading root, like the barberry and the sumach. Under favorable conditions it is a very handsome tree, seldom rising above twelve or fifteen feet, but displaying a round head, and covered in its season with a profusion of flowers, unfolding their beautiful pyramidal clusters regularly on the last week in May. The color of these flowers is perfectly unique, having given the name by which painters distinguish one of their most important tints. The foliage of this tree is not remarkable, except for the regular heart shape of the leaves. It displays no tints in the autumn, but falls from the tree while its verdure remains untarnished.

The Lilac is still cultivated and prized in all our country villages. But its praise is seldom spoken in these days, for Fashion, who refuses to acknowledge any beauty in what is common, discarded this tree as soon as it became domesticated in humble cottage gardens. Even the rose would long ago have been degraded from its ancient honors by this vulgar arbiter of taste, if it had not been multiplied into hundreds of varieties, permitting one after another to take its turn in monopolizing to itself those praises which are due to the primitive rose.

THE BARBERRY.

All the inhabitants of New England are familiar with the common Barberry, one of those humble objects of the landscape that possess great merit with little celebrity. It is allied in picturesque scenery with the whortleberry and the bramble. We see it in hilly pastures, upon soils less primitive than those occupied by the vaccinium, though it is not uncommon as an under-shrub in many of our half-wooded lands. I have not yet been able to obtain a definite idea of the nature of those qualities that entitle a plant to the praises of florists and landscape gardeners, since we find them admiring the ugly mahonia more than the common Barberry, and the glutinous and awkward rose-acacia more than the common locust. The praises of the Barberry have not been spoken; but if our landscape were deprived of this shrub, half the beauty of our scenery would be wanting in many places. Its flowers hanging from every spray in golden racemes, arranged all along in the axils of the leaves from the junction of the small branches to their extremities, always attract attention. But though elegant and graceful, they are not so conspicuous as the scarlet fruit in autumn. There is not in our fields a more beautiful shrub in October, when our rude New England hills gleam with frequent clumps of them, following the courses of the loose stone walls and the borders of rustic lanes. Even after it is stripped of its fruit, the pale red tints of its foliage render it still an attractive object in the landscape.

THE MISSOURI CURRANT.

AMONG the flowering shrubs which are universally admired for the fragrance and beauty of their early blossoms, the Missouri Currant deserves more than a passing mention. Though introduced into New England since the beginning of the present century, it has become a universal favorite in our gardens, where it is cultivated chiefly for the agreeable odor of its flowers, resembling that of cloves, and penetrating the air on all still days in May. This shrub has a small leaf with irregular pointed lobes, turning to a pale crimson in autumn. The flowers are in small racemes like those of the common garden currant, but brighter in their hues, which are of a golden yellow, and producing only a few large berries of a pure shining black. This species is chiefly prized for its flowers, and is not cultivated for its fruit.

THE CEANOTHUS, OR JERSEY TEA.

THE Ceanothus was formerly well known to the people of the United States under the name of Jersey Tea. Its leaves were extensively used as an imitation tea during the Revolution. They seem to possess no decided medicinal qualities, being somewhat astringent, slightly bitter, but not aromatic. It has been learned from experience that the aromatic plants, by constant use as teas, will pall upon the appetite, and injuriously affect digestion; while those which are slightly bitter, but wanting in aroma, like the China tea plant, may be used without seriously affecting the health for an indefinite space of time. I believe it may also be stated as a maxim, that those plants whose properties are sufficiently active to be used as medicines have never been long employed by any people as substitutes for tea.

The flowers of the Ceanothus are white, in full and elegant clusters, without any formality of shape, having a downy appearance, always attracting attention, not so much by their beauty as by their delicacy and their profusion. This plant is abundant in New England, flowering in June on the borders of dry woods.

FOLIAGE.

Foliage is the most conspicuous of the minute productions of nature. To the leaves of trees we look, not only for the gratification of our sense of beauty, but as the chief source of grateful shade and of the general charms of summer. They are the pride of trees no less than their flowers, and the cause of healthful freshness in the atmosphere. They afford concealment to small birds and quadrupeds, they give color to the woods, and yield constant pleasure to the sight without any weariness. It is remarkable that we always trace with delight the forms of leaves in other objects of nature, — in the frostwork on our windows, in the lichens that cover the rocks in the forest, in the figures on a butterfly's wing. Especially in art do we admire the imitation of foliage. It is, indeed, the source of half the beauty of this earth; for it constitutes the verdure of field and lawn, as well as of woods. Flowers are partial in their distribution, but foliage is universal, and is the material with which nature displays countless forms of beauty, from the small acicular leaves of the delicate heath plant, to the broad pennons of the banana, that float like banners over the hut of the negro.

With the putting forth of leaves we associate the most cheerful and delightful of seasons. In their plaited and half-unfolded condition and in their lighter hues we behold the revival of spring, and in their full development and perfected verdure the wealth, the ripeness, and the joyful fruition of summer. The different colors they assume are indeed the true dials of the year; pale shades of all denote its vernal opening; dark and uniform shades of green mark the summer; and those of gold, crimson and russet the autumn; so that by the leaves alone we might determine the month of the year. They form a delightful ground-work both for fruit and for flowers, harmonizing with each and making no discord with any hues of vegetation. If we consider leaves only as individual objects, they will not compare with flowers either in beauty of form or color. A single leaf seldom attracts a great deal of attention; but leaves in the aggregate are so important a part of the beauty of Nature, that she would not possess any great attraction for the sight without them. A cactus, though admired as a curiosity, and as the parent of magnificent flowers, is on account of its leafless habit but a miserable object; and we can imagine how forlorn must be the scenery of those Peruvian regions where the different species of cactus are the principal forms of vegetation.

It is very general to admire foliage in proportion as it is dense and capable of affording an impenetrable shade; but however desirable this may be to yield us a pleasant retreat on a summer noon, the beauty of a tree is not much improved by this quality. At a distance it presents a lumpish and uniform mass, with but little character; while a tree with moderately thin foliage, so thin as to be penetrated by the flickering sunshine, often discovers a great deal of character, by permitting the forms of the branches to be traced through its shadows. When I sit under a tree, I want to see the blue sky faintly glimmering through the leaves, and to view their forms on its clear surface when I look upwards. I would dispense with a profusion of shade, if it could be obtained only by shutting these things out from observation. Hence I always feel a sensation of gladness when rambling in a birchen grove, in which the small thin foliage and airy spray of the trees permit the sun and shade to meet and mingle playfully around my path.

The lumpish character of the foliage of large-leaved trees, like the tulip and magnolia, is perceptible at almost any distance, causing them to appear like green blots upon the landscape. The small-leaved trees, on the contrary, exhibit a certain neatness of spray, which immediately affects the eye with a sensation of beauty. This appearance is beautifully exemplified in the beech. Some of the large-leaved trees, however, possess a kind of formality that renders them very attractive. Such is the horse-chestnut, that spreads out its broad palmate leaves with their tips slightly drooping, like so many parasols held one above another. People have learned to admire large and broad foliage from descriptions of the immense size of tropical leaves, and by associating them with the romance of a voluptuous climate. The long pennon-like leaves of the banana and the wide fronds of the fan palm naturally excite the imagination of the inhabitant of the North.

The form of leaves, no less than their size, has a great share in their general effects, even when viewed from a distant point, where their outlines cannot be discriminated. If they are deeply cleft, like those of the river maple and the scarlet oak, or finely pinnate, like those of the locust and the mountain ash, we perceive a light, feathery appearance in the whole mass, before we are near enough to distinguish the form of individual leaves. This quality is apparent in the honey locust as far off as the tree can be identified. Hence the forms of leaves do not produce all their effect upon a near view; but in ornamental designs in the fine arts the delineations of foliage alone are considered. In the tracery of fenestral architecture, leaves are a very general and favorite ornament;

and in photographic pictures of single leaves, the beauty of their outlines becomes more evident than in nature.

The most remarkable quality of foliage is color; and all will admit that green is the only color that would not produce weariness and final disgust. Omitting what may be said of autumn tints, the different shades of green in the forest, both while the foliage is ripening and after its maturity, constitute a very important distinction of individuals and species. Pure green is rarely found in any kind, except in its early stage of ripeness. The foliage of trees, when fully matured, is slightly tinged with brown or russet, and on the under side with white or blue. Painters, therefore, seldom use unalloyed green in their foliage; for even if they would represent its appearance in early summer, when its verdure is nearly pure, the effects of sunshine and shade upon the green forest can be produced only by a liberal mixture of the warm tints of orange and yellow when the sunshine falls upon it, and of purple and violet when it is in shadow.

If I were to select an example of what seems to me the purest green of vegetation, I should point to grass when smoothly shorn, as in a well-dressed lawn, so that the leaf only remains. By comparing the verdure of different trees with this example, we shall find it generally of a darker shade and inferior purity. The only trees of our soil that seem to me lighter, when in leaf, than grass, are the plane and the catalpa. We must observe trees on a cloudy day to distinguish the different shades of their foliage with precision. In such a state of the atmosphere they are all equally favored by the light; while, if the sun shines upon them, their verdure is modified according to the direction in which it is viewed.

That kind of foliage to which the epithet "silver" is usually applied is a very general favorite; but it is admired only because it is rare. I cannot believe, if the two kinds were equally common, that the silver leaf would be preferred to the green; for this is the color that affords the most enduring satisfaction. The white poplar is the most remarkable example of silver foliage. The river maple has less of this quality, though it seems to be one of the points for which it is admired. Nature displays but very little variegated foliage among her wild productions, except in the spring and autumn. It is evidently an abnormal habit; hence we find this variegation chiefly in those plants which have been modified by the cultivator's art, and it seldom constitutes a specific mark of distinction.

In our studies of foliage we must not overlook the grasses, which are composed almost entirely of leaves. They contribute as much to the beauty of landscape as the verdure of trees, and collectively more than flowers. We need only a passing thought to convince us how tame and lifeless the landscape would be, though every hill were crowned with flowers, and every tree blossomed with gay colors, if there were no grasses or some kind of herbage to take their place. Hence the superior beauty of Northern landscape compared with the general scenery of tropical regions. There are more individual objects in a Southern land which are curious and beautiful, but its want of green fields soon renders its scenery wearisome.

There is also an interest attached to hills and meadows covered with green herbage, and pastured by flocks and herds, that comes from our sympathies and imagination, and causes the verdure of grass, when outspread upon their surface, to possess a moral or relative beauty displayed by few other natural objects. There is nothing else in landscape to be compared with it, and nearly all outdoor scenes would be cold and insipid without it. It expresses the fertility of the soil; it tells of gentle showers that have not been wanting; and it becomes thereby the symbol of providential care, the sign of pastoral abundance and rural prosperity. We find the grasses only where nature has made the greatest provision for the comfort and happiness of man and animals. All the beauties and bounties of springtime and harvest gather round them; the dews of morning glisten upon them like stars in the heavens; the flowers are sprinkled upon them like gems in beautiful tapestry; the little brooks ripple through them with sounds that are always cheerful, and flash in the sunlight as they leap over their bending blades. The merry multitudes of the insect race gain from them shelter and subsistence, and send up an unceasing chorus of merry voices from their verdure, which is a beautiful counterpart of the blue of heaven.

It may be truly said that no splendor of flowers or of the foliage of trees would make amends for the absence of grass. Distant hills and plains may be made beautiful by trees alone; but all near grounds require this velvety covering to render them grateful to the sight or interesting to the mind. This is the picturesque view of the subject; but in the eyes of a botanist grass is almost infinite in its attractions. In every field or pasture that offers its tender blades to the grazing herds, there are multitudes of species, beside the thousands of herbs and flowers and ferns and mosses which are always blended with them, and assist in composing their verdure. What seems to the eyes of a child a mere uniform mass of green is an assemblage of different species that would afford study for a lifetime. Grasses, though minute objects, are vast in their assemblages; but if we reflect on the phenomena of nature, we shall not consider the least thing any less admirable than the greatest. The same amount of wonderful mechanism is indicated in a spear of herdsgrass as in the bamboo that exceeds in height the trees of our forest; and the little cascade that falls over the pebbles in our footpath is as admirable to one who regards it as evincing the power of nature, as the Falls of Niagara.

THE TUPELO.

THE old town of Beverly, which was a part of Salem during the era of witchcraft, abounds, like other townships on the northern coast of Massachusetts Bay, in rugged and romantic scenery. On one of the bald hills of this town, a pond fed by a spring near the top of the hill served as a watering-place for the flocks that were pastured there. The only tree on this elevation of bare granite, interspersed with little meadows of thin soil, covered with sweet-fern and whortleberry-bushes, stood on the brink of this pond. It was an ancient Tupelo, and attracted the attention of every visitor by the singular manner in which it spread its long branches in a crooked and horizontal direction over this emerald pool. It became the wonder of all that the tree should adopt such an eccentricity of habit, hardly showing a single branch on the land side, and bending over the water like an angler sitting at his task. It was evident that it had never been trimmed into this shape by artificial means. Many people, therefore, believed that its grotesque appearance had some connection with witchcraft, and that the witches who were hanged upon it had caused all the branches to wither and fall on the side that held the victims.

This tree has, I believe, no representative on the old continent; and though there are several species in the United States, only one is found in New England. Here it is one of the most remarkable trees as a picturesque object in landscape. Indeed, there is no other tree, not excepting the oak, that will compare with it in certain eccentricities of habit. It has received a variety of names in different parts of the country, being called "Swamp Hornbeam," from the toughness of its wood; "Umbrella Tree," from a peculiar habit of some individuals to become flattened and slightly convex at the top. Among our country people it is known as the "Wild Pear," from a fancied resemblance between its foliage and that of the common pear-tree. The resemblance seems to consist only in the size and gloss of its leaves. In the Middle and Southern States it is called the "Sour Gum," to distinguish it from the "Sweet Gum," or *Liquidambar*. The name of Tupelo was given it by the aboriginal inhabitants.

The shapes assumed by the Tupelo are exceedingly grotesque, though it is frequently as regular in its growth as our most symmetrical trees. It is sometimes quite erect, extending its branches horizontally and pretty equally on all sides, but generally forming a more or less flattened top. More frequently the Tupelo displays no symmetry of any kind, extending its branches mostly on one side, and often putting forth two or three branches greatly beyond all the others. Many of these are considerably twisted, inclining downward from a horizontal position, not with a curve like those of the elm, but straight, like those of the spruce, though without any of its formality. The spray is very different from that of other trees. Every important branch is covered all round, at top, bottom, and sides, with short twigs, at right angles with the branch. Some of the swamp oaks resemble the Tupelo in fantastic shape, but they never have a flattened top.

The Tupelo is the very opposite of the ash in its general characters; the one is precisely regular in its habits, the other eccentric and grotesque. The leaves and small branches of the ash are opposite, those of the Tupelo alternate; the one has a coarse, the other a finely divided spray: so that there are no two trees of the forest so entirely unlike. It is remarkable that an isolated situation, which is favorable to symmetry and good proportions in other trees, increases the specific peculiarities of the Tupelo. If it has stood alone and sent forth its branches without restraint, it then displays the most grotesque irregularity, showing that its normal habit of growth is eccentric.

The foliage of the Tupelo is remarkable for its fine glossy verdure. The leaves are oval, narrowing toward the stem and rounded at the extremity. The flowers are greenish and inconspicuous, borne in minute umbels on the end of a long peduncle. They produce small berries of a deep blue color, containing a hard stone. This tree is one of the brightest ornaments of our forest in autumn; the fine green color of its foliage attracts our attention in summer, and in winter its grotesque forms, rising out of the shallow meres, yield a romantic interest to these solitary places. It is not well adapted to dressed grounds, but harmonizes only with rude, desolate, and wild scenery.

THE HORNBEAM.

The Hornbeams, of which in New England there are two species belonging to a different genus, are small trees, rather elegant in their shape, and remarkable for the toughness and hardness of their wood. The American Hornbeam, or Blue Beech, is distinguished by its fluted trunk, which, as Emerson describes it, "is a short irregular pillar, not unlike the massive reeded columns of Egyptian architecture, with projecting ridges, which run down from each side of the lower branches. The branches are irregular, waving or crooked, going out at various but large angles, and usually from a low point on its trunk." Old Gerard remarks concerning the English Hornbeam: "The wood or timber is better for arrows and shafts, pulleys for mills, and such like devices, than elm or witch-hazel; for in time it waxeth so hard that the toughness and hardness of it may rather be compared to horn than to wood; and therefore it was called Hornbeam."

The foliage of the American Hornbeam resembles that of black birch, neatly corrugated, of a delicate verdure in summer, and assuming a fine tint of varying crimson and scarlet in the autumn. The name of Blue Beech was applied to it from the similarity of its branches to the common beech-tree, while their surface is bluish instead of an ashen color. Though existing in every part of the country, it is not abundant anywhere, and is not in any tract of woodland the principal timber. It is most conspicuous on the borders of woods, by the sides of roads lately constructed. The scarcity of trees of this species near old roadsides has been caused by the value of their timber, which is cut for mechanical purposes wherever it may be found. The wood of this tree is used for levers, for the spokes of wheels, and for nearly all other purposes which require extreme hardness of the material used.

THE HOP HORNBEAM.

The Hop Hornbeam is a very different tree from the one just described, resembling it only in the toughness of its wood, whence the name of Lever-Wood has been very generally applied to it. This tree is rarely seen by the wayside. Those only know it whose occupation has led them to seek it for its service in the arts, or those who have examined it in their botanical rambles. It is a small tree, that affects the habit of the elm in its general appearance, of the birch in its inflorescence, and of the beech in the upward tendency of its small branches. It is so much like the elm in the style of its foliage, in the fine division and length of its slender spray, and in the color and appearance of its bark, that it might easily be mistaken for a small elm, without any of its drooping habit. It does not, like the elm, however, break into any eccentric modes of growth. A striking peculiarity of this tree is the multitude of hop-like capsular heads that contain the seeds.

INSECURITY OF OUR FORESTS.

THE American continent is so vast, and so large a part of it is still covered with wood, that men are not ready to believe there is any danger of exterminating its forests. Supposing them to be inexhaustible, they are entirely indiscriminate in their method of clearing them, and treat them as if they were of no importance further than they subserve the present wants of the community. They are either reckless or ignorant of their indispensable uses in the economy of nature, and seem purposely to shut their eyes to facts and principles in relation to them which are well known to men of science. Our people look upon the forests as valuable only so far as they supply material for the arts and for fuel, for the construction of houses, ships, and public works; and as there is not much danger of immediately exhausting the supplies for these purposes, the public mind remains quiet, while certain operations are going forward which, if not soon checked by some very powerful restraint, will, before the lapse of another century, reduce half this wide continent to a desert. The science of vegetable meteorology deserves more consideration than it has yet received from our professors of learning. This, if fully explained, would teach men some of the fearful consequences that would ensue if a country were entirely disrobed of its forests, and their relations to birds, insects, and quadrupeds would explain the impossibility of ever restoring them. Man has the power, which, if exercised without regard to the laws of nature, may, at no very distant period, render this earth uninhabitable by man. In his eagerness to improve his present condition, and his senseless grasp for immediate advantages, he may disqualify the earth for a human abode.

This matter has been strangely overlooked by legislators in the several States, though frequently discussed by naturalists and philosophical writers. In spite of the warnings the people have received from learned men, very little thought has been given to the subject. How few persons suspect that in less than a century the greatest affliction this country is doomed to suffer may be caused by the destruction of its forests! Springs once full all the year will be dry every summer and autumn; small rivers will desert their channels; once profitable mill-privileges will cease to be of any value; every shower will produce inundations; every summer will be subject to pernicious droughts. The preservation of the forests in a certain ratio over our whole territory ought to be the subject of immediate legislation in all the States. It is not a part of the plan of this work, however, to treat of woods as a subject of political economy, but rather to prompt our wise men to protect them by statute, by showing our dependence on them for our existence.

It has been said that the intelligence of an educated and civilized community like our own ought to save the country from this evil. But it is our civilization that has created the very danger that threatens us. A country, while it remains in the possession of barbarians, is never disforested. It is a false assurance that the general intelligence of the community will secure them from this danger, unless they have studied the causes of it. A literary and even a scientific education, as popularly conducted, does not imply any great amount of this kind of knowledge. The intelligence of our people would undoubtedly prepare them to understand the subject when explained to them by some one who has made it his special study; but reading does not acquaint a person with facts contained only in books which he never reads, though his habit of reading only for amusement may keep him ignorant of many things which he would otherwise learn from observation. The subject of this essay is not sufficiently exciting to obtain a hearing from the public in a lecture-room. Every avenue of popular information is so greatly obstructed by objects designed only to afford amusement, that science and philosophy, save those branches which some eloquent work has rendered fashionable, have but very little chance to be heard. Even among our literary classes, if you speak of trees and woods, there is only an occasional individual of eccentric habits who seems capable of taking any other than an æsthetic view of their relations to human wants.

But it will be said, if a liberal education does not supply men with the right kind of knowledge on this point, certainly our *practical men* will understand it. They, I admit, would see at once how much money could be made by cutting down all the trees in any given tract of forest; but they are not the men to be consulted respecting the advantage of any scheme that does not promise to be a profitable investment of capital. Our practical men are the very individuals from whose venal hands it is necessary to protect our forests by legislation. In France, where great evils have followed the destruction of woods, laws have been enacted for restoring and preserving them in certain situations. These laws, however, originated, not with practical men, but with Napoleon III., who obtained his views from men of science. Our people have less knowledge of this subject than the Europeans, who have been compelled to study it by the presence of evils which the Americans are just beginning to experience.

The sentiment of the American public seems to have been excited in favor of trees individually considered, rather than forests. People look upon trees as their friends; and more indignation is generally caused by the felling of a single large tree standing in an open field or by the roadside, than by the destruction of whole acres of woods. Our love of trees is a sort of passion; but we need yet to learn that a wood on a steep hillside is of more importance than as many standards as there are trees in the same wood, scattered upon a plain. This æsthetic sentiment seems to be the only conservative principle that has yet produced any considerable effect in preserving trees and groves. It often extends to groups of trees, and sometimes to large assemblages, especially on estates which have remained through several generations in the possession of one family. But generally the avarice or the necessity of our farmers has been more powerful to devastate, than the taste and sentiment of others to preserve our woods.

I have long been persuaded that, unless the governments of the several States should make this a subject of special legislation, the security of our forests must depend on men of large property in land. Men of wealth, if not learned, are generally in communication with men of learning, from whom they may obtain a knowledge of vegetable meteorology, and not being obliged, by pecuniary necessity, to cut down their woods, will, from a sense of their importance in the economy of nature, become their preservers. The wealth and taste of certain families in every town and village will save a great many trees, groves, and fragments of forest. But if our law-makers neglect to legislate for this end, we must look to the possessors of immense estates, the lords of whole townships, for the preservation of any large tracts of forest.

There is a sentimental theory of political economy that

condemns large estates, which, if divided into small farms, would support a greater number of human beings. Indeed, the question is very difficult to answer, how large a proportion of the territory of any country may be kept in forest consistently with the greatest amount of agricultural prosperity. But he who believes that every acre of waste land is so much drawback upon national wealth must have very imperfect views of nature's economy. Even if our continent were circumscribed within bounds as narrow as those of Great Britain's isle, the woods ought to be preserved to a certain extent, though they might check the increase of our population. The superfluous lands of the British nobility have saved their country from many evils that could not have been foreseen when their estates were originally divided. The very selfishness of princes and lords has prevented the extirpation of European forests. If, two centuries ago, England had been parcelled out to the people in farms of one hundred acres, there would hardly be a tree remaining at the present time, certainly not a forest in the whole island.

To assist in calling attention to the importance of our forests, I have devoted a considerable number of these essays to the science of vegetable meteorology. I shall treat, under its several heads, of the uses of trees in preserving a general fulness of streams, and an equal supply of moisture to all parts of the surface ; for sustaining the vitality of the atmosphere, and for charging it with vapor, thereby increasing the frequency of showers and preventing long-continued droughts. Considering them also as electric agents, I shall mark the importance of a certain disposition of them to prevent showers from being wasted upon the ocean and large inland collections of water. I shall speak of their relations to temperature and climate, to show in what manner the clearing of the forest may ameliorate, and how, on the other hand, it may ruin, the climate of any country, whether of large or small extent. It will appear that even the soil in many situations has been actually created by the forest that stands upon it, and that it can only be preserved by its continuance. Lastly, I shall prove that the woods in their wild state, and with their undergrowth, are the cause of preserving our fields and gardens from the over-multiplication of insects, by affording a harbor to the birds, without whose services, in the economy of nature, the human race would become extinct.

ORCHARD TREES.

THE orchard trees, though but few of them are indigenous, constitute one of the most important groups, considered as objects of beauty, to say nothing of their utility. The most of this class of trees belong to the natural order of rosaceous plants, among which are some of the fairest ornaments of Northern climes. Such are the cherry, the peach, the apple, the pear, also the mountain ash and its allied species down to the mespilus and hawthorn. These trees are suggestive of the farm and its pleasant appurtenances, rather than of rude nature; but so closely allied is Nature to the farm, when under the care of a simple tiller of the soil, and unbedizened by taste, that its accompaniments seem a rightful part of her domain. The simplicity of the rustic farm is in consonance with the fresh, glowing charms of Nature herself. A row of apple-trees overshadowing the wayside forms an arbor in which the rural deities might revel as in their own sylvan retreats; and Nature wears a more charming appearance, when to her own rude costume she adds a wreath twined by the rosy fingers of Pomona.

The flowers of the orchard trees are invariably white or crimson, or different shades of these two colors combined. Those of the cherry-tree and the plum-tree are constantly white; those of the pear-tree are also white, with brown or purple anthers; those of the peach and apricot are crimson; those of the apple-tree and quince-tree, when half expanded, are crimson, changing to white or blush-color as they expand. The colors of the hawthorn vary, according to their species, which are numerous, from white to pure crimson. Only a few of the orchard trees have been cultivated for their flowers alone; among these we find a species of cherry with double flowers, and a double-flowering almond, which are common in flower-beds. The Virginia crab-apple is also planted for the fragrance and beauty of its flowers; and if the Siberian species had no material value, it would be cultivated for the beauty of its fruit.

As I have frequently remarked, Nature is not lavish of those forms and hues that constitute pure organic beauty. She displays them very sparingly under ordinary circumstances, that we may not be wearied by their stimulus, and thereby lose our susceptibility to agreeable impressions from homely objects. But at certain times and during very short periods she seems to exert all her powers to fascinate the senses. It is when in these moods that she wreathes the trees with flowers for a short time in the spring, and just before the coming of winter illumines the forest with colors as beautiful as they are evanescent.

The APPLE-TREE was one of the first trees planted by the original settlers of New England, who could not in the wilderness raise those fruits that require the skill of the gardener. This tree is indigenous in all parts of Europe, Northern Asia, and North America. On this continent are found two native species, of which the Virginia Crab is the only important one. This tree bears a small green fruit, agreeable, odoriferous, and intensely acid; but our attention is chiefly attracted by its rose-colored flowers, that perfume the whole atmosphere with a sweetness not surpassed by that of the rose. Nothing in the world can exceed the purity of this fragrance, which, in connection with its beautiful flowers, borne in large clusters, render it the admiration of all. The lover of nature is delighted to find this species in a perfectly unsophisticated state, and unimproved by culture, which always tends to insipidity. The Druids paid great reverence to the apple-tree, because the mistletoe grew upon it. In our own fields it is free from this parasite, which is not found on the western continent above the latitude of Virginia.

The apple-tree bears some resemblance to the oak in its general outlines, displaying, though inferior in size, more sturdiness than grace. A standard apple-tree commonly resembles a hemisphere, often in diameter exceeding its own height. This shape might be caused by training; but the gardener, by cutting off certain branches, does not change the tendency of the tree to assume its normal shape. The foliage of the apple-tree is rather coarse, stiff, and inelegant, and deficient in purity of verdure, being after it is fully developed of a dusky green, and without tints when ripened, save what may be termed accidental. There is, nevertheless, a certain kind of beauty in an old apple-tree which is seen in no other of the orchard trees, rendering it a very picturesque object in rustic scenery.

The PEAR-TREE is taller than the apple-tree, assuming an imperfectly pyramidal shape. Its branches have not the horizontal tendency of the latter; but when growing singly as a standard it greatly surpasses it in dimensions, and many individuals of a former age, that have escaped the axe of horticultural improvement, are noble standards, and of no inferior merit as shade-trees. The foliage of the pear-tree displays some of the tremulous habit of the aspen, owing to the length and slenderness of its leaf-stems. It has, moreover, a gloss that distinguishes it from that of the apple-tree; it is also less stubborn in retaining its verdure, and partially tinted in autumn. The pear-trees which have been raised within the last thirty years are mostly dwarfed, and seldom display their normal shape. They are small, with straggling branches, and unworthy of consideration in a treatise of this kind. The old standards, still occasionally seen in pastures and fallow lands, are the only ones that affect the beauty of landscape. I have mentioned several points in which the pear-tree surpasses the apple-tree as a beautiful and stately object; but its fruit will bear no comparison in beauty with that of the apple-tree, which produces a greater variety of beautiful fruit than any other tree that is known.

The QUINCE-TREE, though inferior in size, and not prospering very well on the soil of New England, which is rather too cold for it, deserves a passing remark. In botanical characters it bears more resemblance to the pear than to the apple. The fruit has the same tender and mucilaginous core; the seeds are not enclosed in a dry hull, like those of the apple; and the pulp of the quince, like that of the pear, is granulated, while that of the apple displays in its texture a finer and firmer organization. I may add the well-known fact that the pear may be grafted upon a quince stock, while no such union can be effected between the apple and the quince, or the apple and the pear. The quince-tree makes a very elegant appearance, both when covered with its large white and crimson-stained flowers, and when laden with its golden Hesperian fruit.

The PLUM-TREE, in connection with the orchard, hardly deserves mention; but there are two indigenous species which in some places are conspicuous objects in our fields. The beach-plum requires no description. It is a low shrub, very common on many parts of the New England coast and on the islands around it. There is nothing remarkable in its appearance or in the beauty of its fruit, which is of a dark-blue color and about the size of damsons. The other species is a tree of considerable size, which is very beautiful when covered with its ripe scarlet berries. In the State of Maine they are called "plumgranates," and are very generally used for culinary purposes.

The PEACH-TREE, of all the tenants of the garden and orchard, is the most beautiful when in flower, varying in the color of its bloom from a delicate blush to a light crimson. As it puts forth its flowers before the leaves, the tree presents to view the likeness of a magnificent bouquet. When covering many acres of ground, nothing in nature can surpass it in splendor, flowering, as it does, sooner than almost any other tree. Even in New England, where these trees are now seen only in occasional groups, they constitute an important object in the landscape, when in flower. Few persons are aware how much interest the peach-tree adds to the landscape in early spring, by its suggestions as well as its beauty. Since the changeableness of our winter and the harshness of our spring weather have been aggravated by the destruction of our Northern forests, the peach-tree is so liable to perish that its cultivation has been neglected, and trees of this species are now very scarce in New England, except in the gardens of wealthy men. We no longer meet them as formerly in our journeyings through rustic farms, when they were interspersed among apple-trees, adorning every by-way in the country.

WAYSIDE SHRUBBERY.

There are some persons in the world whose ideas of beauty run almost entirely into waving lines, smoothness, and rotundity. They cannot bear to see anything in landscape that does not convey the sentiment of costly dressing and ornamentation. In their sight nothing is so beautiful as a well-trimmed hedge-row, or a nicely painted fence. They abhor any appearance of rudeness about their grounds, or anything that is not an evidence of wealth and "æsthetic culture"; believing that the more completely Nature is subdued by Art, the more credit she reflects upon her owner! They regard Nature as they do their horse, and believe, if she can be made to look sleek and plump, they have fully carried out one of the beneficent designs of Providence, whom they regard as the great teacher of æsthetics!

Unfortunate are the picturesque old roads that fall under the management of this class of men, when employed as surveyors of the highways. In their view, no spontaneous production of nature should remain there, except the gravel. Not even a tree must be tolerated, unless it was planted there by human hands. All wild growths of shrubbery are condemned to perish. If a by-road be embroidered with a charming variety of native shrubs and herbaceous plants, at one season adorned with flowers, at another with fruits, and at all times with foliage, they send the plough directly into this mass, to carry out their ideas of neatness and smoothness, and to expel Nature as if she were a Gorgon or a Hydra.

The passers on this by-road formerly wended their way along a footpath, through its various shrubbery; and the children of the village, as they went to school and returned, would often linger here to gather flowers and fruit, sitting down upon some green tussock, under the shady protection of half-grown trees, which had come up without planting. They watched the viburnum, with its circular cymes of white flowers, succeeded by blue, white, and purple berries; the wild roses that clustered there in June, and the glycine that festooned the thickets with dark flowers in August. They admired the charming negligence of these growths, some with upright stems supporting the twining convolvulus, interwoven with the dark-blue flowers of the woody nightshade, and others climbing overhead and forming an arbor for a summer noonday. The surveyor and his gang have spoiled the footpath, and destroyed the bushes with their flowers and fruit; and children no less than birds lament this destruction of their pleasant wayside haunts. Ever since my boyhood have these vandals of the roads been deservedly cursed as the despoilers of nature, and the clumsy agents of tasteful imposture.

There is another class of despoilers who pursue their operations as private citizens. They are generally "model farmers," — men who think that nature should be made subservient to labor, and labor to capital. If you stroll along by the estates of these industrious vandals, you will be struck with the baldness and nakedness of the borders of their fields. Not a shrub nor a vine can with impunity lift its head above the ground on either side of their fences, and a squirrel that should venture near them would be hunted like an adder. We may distinguish the possessions of these model farmers by observing, as we pass by, their singular blankness, such as you observe in the face of an overfed idiot. Their

Nature is a young damsel with her hair tied up in knots and papers, as distinguished from one whose tresses hang down her neck in careless freedom.

They will tell you that wild shrubbery harbors vermin, and that its intricacy affords them shelter, which is not provided by a formal hedge-row. But if it harbors insects, it protects also the birds that feed upon them; if it causes the multiplication of small quadrupeds, it supplies also the mast that sustains them. In proportion as the thicket has been eradicated from the borders of fields and waysides, the insects that destroy our crops have increased; for the birds that once found protection within it have fled to distant places, and left the insects to commit their ravages unmolested.

There is a certain kind of beauty in high cultivation; there is still more in neatness and simplicity; there is even a sort of relative beauty in baldness, when it is plainly necessary to make a free and convenient passage for a constantly moving crowd. There is no man who cannot appreciate the beauty of a well-tilled farm, nor is there one who would prefer rubbish and litter to neatly dressed paths and borders. But if it be a question whether the perfect smoothness and baldness of a gravelled walk, weeded of every bush and herb, is more interesting than the rustic negligence of another walk, covered with a variety of shrubs and vines, and intersected by a footpath worn by the feet of men and animals, there are but few, even among the most sordid, who would not prefer the neglected pathway. For is it nothing to us that the singing-birds should find a bushy knoll to nestle in, or a leafy perch to rest upon when they sing to the passing traveller? Is it nothing to us that we may gather a few violets under a hazel-bush for the little child we lead by the hand? Is it nothing to the young maiden that she can loiter by the roadside, in quest of wild-flowers, instead of roaming in distant fields, where she dares not venture unprotected?

On those by-roads where there is but little passing, all kinds of native shrubs are more valuable, as well as more beautiful, than anything that could be put in their place. Especially in the borders of fields near the town are these spontaneous growths, with their grassy turf embossed with wild-flowers, needful for the protection of birds that live in shrubbery, and not in trees, and will not accept the bushes of the garden because they have no tangled undergrowth. Insects, in the different stages of their existence, multiply with the increase of tillage; for every fertilizer that is mixed with the soil renders it more productive of vermin. Birds, which are the natural checks to the over-multiplication of insects, would become more numerous in proportion to the increased supply of their insect food, if there was a harbor for them in the vicinity. A great number of small birds, all of which, not excepting the granivorous species, feed their young with larvæ, are exiled by the want of border shrubbery. The catbird, an inveterate consumer of insects, second only to the robin in usefulness to the farmer, will become very familiar, and build in our gardens, if supplied with a plenty of wild thicket to yield it that seclusion that suits its temper and habits. Birds of every species prefer a certain description of shrubs or trees for a resort; and how great soever their supply of food, if no woods or thickets are near to afford them a harbor and a nursery for their young, they will leave it untasted. Any man who owns an acre of land might gather round it nearly every species of our small birds by a very little sacrifice of space, which is to be filled with the wildings of nature.

A formal clipped hedge-row affords the birds no such shelter nor seclusion. Why an ugly mass of sticks, with a few leaves on the outside, should be preferred to a beautiful growth of wildings, cherished upon a natural sod, is no less wonderful than the "mystery of godliness." As ornaments to the landscape, shrubs are of no secondary importance. While trees afford grandeur to the distant prospect, the beauty of a near prospect is greatly dependent on shrubbery. A rocky and uneven surface, covered by trees alone, would not affect the mind with so many agreeable impressions as when combined with their undergrowth. When we are journeying, the wild shrubs that skirt the waysides, and hang their foliage, fruit, and flowers over the walls and fences, add a beauty and interest to the scenes of our journey not equalled in any respect by the cultivated exotics in the spaded lands of gardens in our suburban towns. In some foreign countries the superstitions of the people cause them to preserve many of these things which are so valuable in ways but little understood. It is the misfortune of our land, that these conservative superstitions which we have rejected are not supplanted by philosophy. In place of it we discover only a barren infidelity to Nature, and a sceptical disregard of her benevolent laws.

The growth of spontaneous shrubbery by the sides of the less frequented roads, and in all those situations where it does not interfere with needful operations, is one of the chief blessings of nature. It is profitable for shade and for shelter, and affords constant pleasure to the sight. Especially on the borders of rustic by-ways it covers the nakedness of the stone wall with foliage and flowers, and produces an abundance of wild fruit for children and birds. Tons of whortleberries would be produced every summer by rustic waysides and the borders of fields, if the sordid land-owner did not destroy the bushes that yield them. I cannot see that a growth of this kind close to the fences would diminish the space that should be occupied by the farmers' crops. Yet were it not for the persistent efforts of Nature, who plants her shrubs with liberal hand in neglected fields and borders, they would long since have been exterminated in all our old townships. It is true that they do not yield any immediate profit to the farmer; but they produce fruit which is a luxury to the children of the neighborhood; they are valuable for the shelter they afford to the birds; they protect the immediate grounds from the winds, even more than trees; and they constitute the most interesting embellishments of a rustic farm.

THE AMERICAN ELM.

I WILL confess that I join in the admiration so generally bestowed upon the American Elm. To me no other tree seems so beautiful or so majestic. It does not exhibit the sturdy ruggedness of the oak; it is not so evidently defiant of wind and tempest. It seems, indeed, to make no outward pretensions of strength. It bends to the breeze which the oak defies, and is more seldom, therefore, broken by the wind. The Elm is especially the wayside tree of New England, and it forms the most remarkable feature of our domestic landscape. If there be in any other section of our land as many, they are individuals mingled with the forest, and are not so frequent by the roadsides. In this part of the country the Elm has been planted and cherished from the earliest period of our history, and the inhabitants have always looked upon it with admiration, and valued it as a landscape ornament above every other species. It is the most drooping of the drooping trees, except the willow, which it surpasses in grandeur and in the variety of its forms.

Though the Elm has never been consecrated by the muse of classic song, or dignified by making a figure in the paintings of the old masters, the native inhabitant of New England associates the varied forms of this tree with all that is delightful in the scenery or memorable in the history of our land. All spacious avenues are bordered with elms, and their magnificent rows are everywhere familiar to his sight. He has seen them extending their broad and benevolent arms over many a hospitable mansion and many a humble cottage, and equally harmonizing with all. They meet his sight in the public grounds of the city with their ample shade and flowing spray; and he beholds them in the clearing, where they were left by the woodman to stand as solitary landmarks of the devastated space. Every year of his life he has seen the beautiful hangbird weave his pensile nest upon the long and flexible branches, secure from the reach of every foe. From its vast dome of branches and foliage he has listened to the songs of the late and early birds, and under its canopy he has witnessed many a scene of rustic amusement.

To a native of New England, therefore, the Elm has a character more nearly approaching that of sacredness than any other tree. Setting aside the pleasure derived from it as an object of material beauty, it reminds him of the familiar scenes of home and the events of his early life. How many a happy assemblage of children and young persons has been gathered under its shade in the sultry noons of summer! How many a young May queen has been crowned under its tasselled roof, when the greensward was just daisied with the early flowers of spring! And how often has the weary traveller rested from his journey under its wide-spreading boughs, and from a state of weariness and vexation, when o'erspent by heat and length of way, subsided into quiet thankfulness and content!

In my own mind the Elm is intimately allied with those old dwelling-houses which were built in the early part of the last century, and form one of the principal remaining features of New England home architecture during that period. They are known by their broad and ample but low-studded rooms, their two stories in front, their numerous windows with small panes, their single chimney in the centre of the roof, that sloped down to one story in the rear, and their general homely appearance, reminding us of the simplicity of life that characterized our people before the Revolution. Their very homeliness is attractive, by leaving the imagination free to dwell upon their interesting suggestions. Not many of these venerable houses are now extant; but whenever we see one, it is almost invariably accompanied by its Elm, standing upon the green open space that slopes down from it in front, waving its long branches in melancholy grandeur above the old homestead, and drooping, as with sorrow, over the infirmities of its old companion of a century.

Early in April the Elm puts forth its flowers, of a dark maroon color, in numerous clusters, fringing the long terminal spray, and filling up the whole space so effectually that the branches can hardly be seen; they appear at the same time with the crimson flowers of the red maple, and give the tree a very sombre appearance. The seeds ripen early, and being small and chaffy are wafted in all directions and carried to great distances by the wind. In the early part of June, soon after the leaves are expanded, the Elm displays the most beauty. At this time only can its verdure be considered brilliant; for the leaf soon fades to a dull green, and displays no tints, except that of a rusty yellow in the autumn. In perfectly healthy elms, standing on a deep soil, the brightness of the foliage is retained to a later period; but the trees near Boston have suffered so much from the ravages of the cankerworm that their health is injured, and their want of vitality is shown by the premature fading and dropping of their foliage.

Nothing can exceed the American Elm in a certain harmonious combination of sturdiness and grace, — two qualities which are seldom united. Along with its superior magnitude, we observe a great length and slenderness of its branches, without anything in the combination that indicates weakness. It is very agreeable to witness the union, under any circumstances, of two interesting or admirable traits of character which are supposed to be incompatible. Hence the complacency we feel when we meet a brave man who is amiable and polite, or a learned man who is neither reserved nor pedantic. A slender vine, supported by a sturdy tree, forms a very agreeable image; not less delightful is that consonance we perceive in a majestic Elm, formed by the union of grandeur with the gracefulness of its own flowing drapery.

The Elm is generally subdivided into several equal branches, diverging from a common centre at a small distance above the ground. The height of this divergence depends on the condition of the tree when it was a seedling, whether it grew in a forest or in an open field; and the angle made by these branches is much wider when it obtained its growth in an isolated situation. The shape of different elms varies more than that of any other known species. It is indeed almost the only tree which may be said to exhibit more than one normal figure, setting aside those variations of form which are the natural effects of youth and age. The American Elm never displays one central shaft to which the branches are subordinate, like the English Elm; or rather, I should say, that when it has only a single shaft it is without any limbs, and is surrounded only with short and slender twigs. This leads me to speak of its normal diversities of shape, which were originally described by Mr. Emerson under several types.

THE DOME.

This is the form which the Elm seems most prone to assume when it stands from the time it was a seedling until it attains its full stature in an open space. It then shows a broad hemispherical head, formed by branches of nearly equal size, issuing chiefly from a common centre, diverging first at a small angle, and gradually spreading outward with a curve that may be traced throughout their length. A considerable number of our roadside elms are specimens more or less imperfect of this normal type.

THE VASE FORM.

One of the most admirable of these different forms is that of the vase. The base is represented by the roots of the tree as they project above the ground, making a sort of pedestal for the trunk. The neck of the vase is the trunk before it is subdivided. The middle of the vase consists of the lower part of the branches as they swell outwards with a graceful curve, then gradually diverge, until they bend over at their extremities and form the lip of the vase by a circle of terminal spray. Perfect specimens of this beautiful form are rare, but in a row or a grove of elms there are always a few individuals that approximate to this type.

THE PARASOL.

The neatest and most beautiful of these forms is the parasol. This variety is seen in those elms which have grown to their full height in the forest, and were left by the woodman in the clearing; for such is the general admiration of this tree, that great numbers of them are left in clearings in all parts of the country. The State of Maine abounds in trees of this form, sending forth almost perpendicularly a number of branches, that spread out rather suddenly at a considerable height, in the shape of an umbrella. Trees of this type have much of that grandeur which is caused by great height and small dimensions, as observed in a palm-tree. A remarkable trait in the character of the Elm is, that, unlike other trees, it seldom loses its beauty, and is often improved in shape, by growing while young in a dense assemblage. It is simply modified into a more slender shape, usually subdivided very near the ground into several branches that diverge but little until they reach the summit of the wood. Other trees, when they have grown in a dense wood, form but a single shaft, without lateral branches.

THE PLUME.

The most singular of the forms assumed by the Elm, and which cannot be regarded as of a normal character, is the plume, caused by some peculiar conditions attending its early growth. The shaft is sometimes double, but usually not divided at all, except into two or three small branches at its very summit. It is perpendicular to near three fourths of its height, and then bends over, like one of the outer branches of a normal-shaped Elm. This whole tree, whether double or single, is covered from the ground to its summit with a dense embroidery of vine-like twigs that cluster round it in all ways, often inverted, as if it were covered with a woody vine. The cause of this form seems to be the removal of the tree into an uncongenial soil, that is too scanty and innutritious to sustain a healthy growth. Yet I have seen some trees of this shape in clearings. They do not seem to be diseased, yet they are evidently in a stunted condition. One of the most remarkable of the plume elms which I have seen stands in the northern part of Danvers, near the point where the Essex Railroad crosses the Ipswich River. I have observed a similar habit of growth in some English elms, but their shaft is always perpendicular.

THE ENGLISH ELM.

The English Elm may be seen on Boston Common, and in front of old mansions in Medford and other ancient towns in Massachusetts. Very few trees of this species, however, have been planted since the Revolution. This royal Elm seems to have lost favor when republicanism took the place of monarchy. Yet in many points the English Elm is superior to the American species. It is not a drooping tree; it resembles the oak in its general form, but surpasses it in height. The trunk is not subdivided; throughout its entire length, the branches are attached to it by wide angles, sometimes spread out in an almost horizontal direction. Selby remarks, that, "in point of magnitude, grandeur of form, and majestic growth, the English Elm has few competitors in the British sylva." In the form of the leaf and spray it closely resembles the American tree; but the leaf is of a brighter green, it comes out several days earlier in the spring, and continues green in the fall a week or ten days after the American elm has become entirely denuded. The same difference, in a less degree, has been observed in the leafing and falling of the leaf of all European trees, compared with their kindred species in the American forest.

ODORS OF VEGETATION.

THE beauty of a summer landscape is greatly enhanced by its alliance with the agreeable odors that constantly emanate from herbs and flowers; for the sight of a grove or woody pasture invariably suggests the idea of fragrance. The rising mists of the valley, tinged with the ruddy hues of dawn, derive interest from their relation to the fragrance of morning. And it may be remarked, on the other hand, that odors are indebted to other charming influences of nature for a great share of their own pleasantness. For nature has so combined all the objects of creation, that they are made to reflect a portion of their own light, beauty, and agreeableness upon each other.

The sense of smelling is not included by philosophers among the intellectual senses, like those of sight and hearing. It chiefly serves the purpose of directing animals to the right selection of the substances they use for food by their agreeable odors, and averting them from such as are noxious by those of an offensive character. This instinct is an unerring guide to the inferior animals among the simple productions of nature. But art is so ingenious in imparting the savor of any agreeable and wholesome substance to others which are injurious, that the sense of smell, even when assisted by taste, is an unsafe guide in the use of artificial preparations. Among natural productions, unmodified by art, the senses of smell and taste are safe guides to all fruits and other substances.

It is not my purpose, however, to discuss this point

physiologically, but to treat of the odors of plants chiefly as the cause of agreeable sensations, and as a sort of picturesque attraction, when we are either rambling in the fields or employed in rural occupations. We perceive characteristic odors in every wood and meadow, by which we recognize their predominant trees, herbage, and shrubbery. Those of an oak wood are very remarkable, and not to be mistaken for any others. They are not aromatic; but they have a freshness more agreeable, perhaps, if we constantly breathed them, than a spicy fragrance. This odor is very similar to that of oak timber when cut and sawed; in one sense, a maritime savor, like that of a ship-yard. To a Briton it is probably a spice of royalty. It comes chiefly from the foliage after it has dropped from the trees; for the fresh green leaves seem to be scentless.

In wet grounds covered with alder, when it is in flower, a very agreeable essence is perceptible in the air; but I have not ascertained its source, whether it comes from the herbage or the shrubbery. It is probably the aroma of its tasselled flowers. I wonder that Darwin, in his "Loves of the Plants," never suggested the idea that the pollen of flowers is guided by these subtle essences to the bosom of its female, when wandering upon the winds. This delicate aroma, perceived when the alder is in flower, is displaced by the more penetrating odor of the azalea in July, and of the clethra in August. The fragrance of these shrubs, combined with that of the myrica and the cranberry-plant, forms the characteristic odor of low grounds, where no stagnant waters are present to mix with it any impurity. It is the primitive odor of the moorlands when covered with their native herbs.

As we leave the meadows and ramble near the hillside, where the native grapevines abound, we perceive another class of odors, still more agreeable, resembling the perfume of mignonette, most perceptible when the vines are in flower. This is the true ambrosia of the gods, — the honey-scent of Mount Hybla. It seems as if nature had infused into the leaf or flower of all plants that bear an agreeable fruit some odor that shall be a reminder of its presence. The scent of the grapevine comes chiefly from its flowers, that of the strawberry-plant from its foliage and fruit. Both leaf and flower of the same plant are seldom fragrant. The flower of the sweetbrier has very little scent compared with that of the common wild rose. The insect, whose services are so valuable to the species, needs not the odor of the flower if it can perceive that of the leaf.

The characteristic odors of the seasons come chiefly from flowers in the spring and early summer, from herbs and foliage in the latter summer, and from the ripened harvest and withered leaves in autumn. Winter is without odors, except those of the forest and seaside. The first aroma that pervades the atmosphere in spring is that of willows and poplars, which are very distinct; the former resembling that of lilacs, the latter more balsamic, and proceeding no less from the glutinous buds than from the flowers. Nature never seems so capricious as when she distributes her odors among the different species of vegetation. Why should the flowers of the elm and the maple be scentless, differing in this respect so notably from other spring flowers? Fragrance is denied them, perhaps as a superfluity, because they bloom and fade before the insect tribes are abroad.

We are all familiar with the scent of flowering orchard trees. It is the incense that May diffuses over the landscape just before her departure. The blossom of linden-trees succeeds, and brings along with it a universal hum of insects, that seem intoxicated with its sweets. From this bloom the bee gathers the choicest honey.

If the linden-tree had no other extraordinary merit, I should preserve it for its unrivalled sweetness. Its fragrant emanations are scattered abroad so widely that not an insect loses a message from its proffered feast of nectar; and the hum of the innumerable hosts of different species attracts our attention as one of the picturesque phenomena of the season.

The true seasonal fragrance of summer is that of new-mown hay, for the air is filled with it during all the time of haymaking. This is indeed the "balm of a thousand flowers"; for though a greater part of the aroma comes from the leaves of clover and different kinds of grasses, the whole is the grateful result of many species with their flowers, when cut down by the scythe. Almost any combination of healthful herbs, when spread out to the sun and wind, after being mowed, will produce an aroma like that of new-mown hay. If you mix with these any considerable quantity of those noxious or innutritious herbs which are not acceptable to cattle, there comes from the mixture a rank herbaceous smell that indicates their presence. Nature is always true to the instincts of her creatures, and sets up no false allurements to tempt them to that which is unhealthful.

To the scent of new-mown hay succeeds that of the grain harvest, — the odor of ripened vegetation. We now mark the difference between the savor of herbs when they are cut down in blossom and after they have ripened their seeds. The odors of summer are more spicy or aromatic, and have more of an intoxicating quality, than those of the harvest. Nature has denied fragrance to the autumnal flowers, except a few that resemble the flowers of spring; such is the graceful neottia, breathing the odor of hyacinths, which is so obscure that it would be overlooked by the insects, amid the host of scentless flowers, if they were not guided by its perfume. Autumn indeed seems niggardly of her gifts to the honey-sipping insects, for the flowers of this season are as destitute of sweetness as of fragrance. The charms of autumn are chiefly for the eye, — of tinted woods and gorgeous flowers, that attract us more by their glowing profusion than by any particular beauty as individual objects.

THE CHERRY-TREE.

AMONG our fruit-trees the Cherry occupies the most conspicuous place, considered with reference either to shade or ornament, surpassing all the others in size and in comeliness of growth. All the species are handsome trees, and some of them are of great stature. They are natives of all countries in the northern temperate zone, but not of any region south of the equator. The three most remarkable species of the family are the common garden Cherry, or Mazard, which is believed to be a native of Asia; the Great Northern Cherry, or Gean, of Europe; and the Black Cherry of the United States.

THE BLACK CHERRY.

The Black Cherry, which is a tree of the first magnitude in favorable regions, is only a middle-sized tree in the New England States. In the South and West, especially on the banks of the Ohio River, it attains a very great size, rising sometimes to one hundred feet, according to Michaux, with a corresponding diameter. It is sensitive to the extremes both of cold and heat, and to an excess either of dryness or moisture. In Maine it is only a small tree, being checked in its growth by the severe Northern winters. Very far south it suffers from the hot and dry summers, but prospers well in the mountainous parts. It forms immense forests in many districts of North America, in company with the honey locust, the black walnut, the red elm, and the oak. It is sufficiently common in New England to constitute an important ingredient of our wood scenery, and though indigenous, it is most abundant in lands which have been modified by cultivation.

This tree differs very obviously in its ramification from the garden cherry, in which the branches are always subordinate to the trunk, and arranged in irregular whorls and stages, one above another, so that, if they were horizontal, they would resemble those of a fir-tree. The Black Cherry tree, on the contrary, is subdivided in such a manner that the main stem cannot easily be traced above the lower junction of the branches, except in those which have grown in a forest. The branches are spread out more loosely, without the least of any arrangement in whorls, and their terminations are longer and smaller. The leaves of the two trees are also widely different: those of the garden cherry are broad, ovate, rough, and serrate; those of the American tree are lanceolate and smooth, and almost as slender as the leaves of the willow. The one bears its flowers and fruit in racemes, the other in round clusters or umbels. The trunk and bark of the two species are similar, both resembling the black birch in the properties of their wood and the outside appearance of their bark. The branches of the Wild Cherry are too straggling and sparse to make a beautiful tree, and the leaves being small and narrow, the whole mass is wanting in depth of shade.

THE CHOKE CHERRY.

When we are rambling in rustic lanes, that lead through rudely cultivated grounds, we frequently meet with groups of tall handsome shrubs, covered in May with a profusion of white flowers, and in August heavily laden with bright scarlet fruit. Such is the Choke Cherry, a small tree with which all are familiar from their frequent disappointment on attempting to eat its fruit. Its promises to the sight are not fulfilled to the taste. Though of an agreeable flavor, it is exceedingly harsh and astringent. This is a more beautiful tree when in flower than the black cherry, though it is generally a mere shrub, never rising above fifteen or twenty feet in height. The racemes, when in flower, are not drooping, as they are when laden with fruit, but stand out at right angles with the branch, completely surrounding it, and giving to every slender twig the appearance of a long white plume. In the eastern part of Massachusetts I have found this species, as well as the black cherry, in old graveyards, — so frequently, indeed, that in my early days these trees were associated with graves, as the Lombardy poplar is with ancient avenues. I suppose their frequency in these places to be caused by the birds dropping the seeds at the foot of the gravestones, where they quickly germinate, and are protected, when growing, by the stone beside them.

The cultivation of the Gean, or Great Northern Cherry of Europe, which was named by Linnæus the bird cherry, is encouraged in Great Britain and on the Continent of Europe for the benefit of the birds, which are regarded as the most important checks to the over-multiplication of insects. The fact, not yet understood in America, that the birds which are the most mischievous as consumers of fruit are the most useful as destroyers of insects, is well known by all the farmers in Europe; and while we destroy the birds to save the fruit, and sometimes cut down the fruit-trees to starve the birds, the Europeans more wisely plant them for their sustenance and accommodation.

THE DREARY AND DESOLATE.

It may be thought somewhat inconsistent with the purpose of these essays to describe the charms of scenery which many regard as disagreeable; for it would hardly be supposed that, while praising a beautiful face, we should dwell with pleasure on its plain or unpleasant features. Yet an expression of sadness, which is not a genuine ingredient of beauty, may excite love by awakening our sympathies; and, to pursue this analogy a little further, we know there is something in the face of certain persons which is superior to beauty, and is the cause why some women, who were not considered beautiful, have inspired the most exalted passion. The pleasure afforded to all imaginative minds by dreary and desolate scenery has its origin in the sentiment of melancholy. This kind of scenery is not identical with simple rudeness; it is more like that which in poetical language is termed weird. We feel the force of the sentiment it awakens when wandering over extensive bald hills, sparsely covered with vegetation, and interspersed with a few trees of gaunt and shaggy appearance, or when traversing wide moorlands, half covered with stagnant waters, and with trees and shrubs blackened by a subsided stream or lake.

We are not always aware how nearly allied are our thoughts and sensations, and may often suppose that the mind is pondering on some intellectual theme or following a metaphysical train of thought, when we are only indulging in the luxury of emotion. And this may explain why a certain vague style of writing, like that of

Renan, if it be really poetical, is highly enjoyed by a class of readers who are inclined to look on all subjects through the colored medium of passion. The poems of Ossian are remarkable for this quality of style. They are tinged with a deep pathos, without relating any incident that acts powerfully upon our sympathy. They are especially distinguished by their power of awakening in the mind the poetic sentiment of desolation. When the bard speaks of the meteors of night that set on the hill before the wanderer, of the faint roaring of distant torrents, of inconstant blasts rushing through the aged oaks, and of the half-enlightened moon that sinks dim and red behind the hill, the beauty of his description depends on its power of exciting those ineffable emotions that flow from sweet, solemn, and melancholy music.

It is from this sentiment of the dreary and desolate that certain peculiar words and images, which have been mostly confined to poetry, derive their forcible expression. Edgar Poe used these forms of speech with singular felicity; and the charm of his poems flows chiefly from his mystic and beautiful euphemisms. The imagery of his poem entitled "Ulalume" produces much of the sensation with which we contemplate a weird or desolate scene in nature. His relation, in the poem, of his wandering with Psyche in the ghoul-haunted woodland of Weir, is full of a certain dreary sentiment of pathos, made still more poetical by its obscurity, like a rude landscape involved in luminous mist. When the skies were ashen and sober, on a gloomy autumnal night, in the misty region down by the dank tarn of Auber, he held a sacred interview with Psyche; and their discourse, without conveying to the mind any clear and intelligible thought, excites very definite sensations of mingled beauty and solemnity. There are intellectual emotions that want the distinctness of thought, and which cannot be so well described by words as by music. But there are words in every language, even in our own unmusical tongue, which are capable, by their sweetness of tone, of powerfully exciting the imagination. Such materials were used with rare skill by this singular and extraordinary genius, who considered the art of using language so as to produce the greatest effect no less worthy of study than the arts of painting and sculpture.

There is nothing positively agreeable in dreary or desolate scenery, yet the sentiment it inspires is associated with a beneficent law of our nature that causes a little pathos and a little melancholy to heighten the pleasures of life. We have all, at certain times, been deeply affected by scenes of dreariness and mystery, and felt from the sweetness and sadness that are blended with them that no merely beautiful scene could awaken the same amount of pleasurable emotion. Who has not felt the charms of a wide solitary plain, of a dark forest, and of the deep ghostly shadows of night on the hills? There is nothing that comes from a mere view of nature that will compare with the luxury of this sentiment, and it seems to me to have been less appreciated by painters than by musicians. I have never seen a picture which equalled the sublimer strains of music, like certain passages in the works of Beethoven, in the power of inspiring these sensations. But the ear is a more emotional organ than the eye, and perhaps a higher order of genius is required in the painter than in the musician to produce equal emotional effects.

We delight to witness the phenomena of Nature under aspects that present her to our imaginations as a gentle sympathizer or seeming partner in our afflictions. Hence autumn is the favorite season of poets, because it emblemizes sorrow, and fills the lap of Nature with dead leaves which she strews over the graves of flowers; and we love to hear the low moaning of winds at this time, when they seem like dirges over the departed beauties of summer. We love the evening twilight and Hesper's melancholy star, because they inspire tender sensations of melancholy, and raise our souls at the same time to the contemplation of infinity. The pale light of the moon gives us intimations of the sympathy of the benign goddess; and while sitting under her light, lovers and mourners, those who rejoice and those who weep, feel the presence of a divinity and an alleviation of those passions that agitate the soul.

The sense of weirdness which we feel when surrounded by certain kinds of dreary landscape intensifies our love of nature. The feelings it inspires are of a spiritual cast, and far above the sensual delights that spring from the sight of dressed grounds and voluptuous gardens. But this kind of scenery may be compared to certain pathetic or solemn strains in music, which, if long continued, would become depressing. When we emerge from such a prospect into that of an opposite character, we feel a pleasant exhilaration which is due to our previous depression. It is like the twilight of morning, that exalts our spirits by blending with the sadness of earth and night some of the inspiring tints of heaven and immortality.

THE SNOWY MESPILUS.

THIS tree, which is conspicuous in the early part of May from its profusion of white flowers in the swamps, is very little known except in Canada and some of the northern provinces of this continent. Yet it is far from being rare, and is one of the most elegant of the small trees in our native forest; being allied to the mountain ash, branching in a similar manner, but exhibiting a neater and more beautiful spray. It is exclusively a Northern tree, and one of the earliest to put forth flowers and leaves after the elm and the red maple. This tree is spread over almost all the northern part of the American continent and the Alleghany Mountains. From its habit of flowering at the time of the annual appearance of the shad in our waters, it is very frequently called the Shad-bush.

The Snowy Mespilus is one of those trees which botanists have described under so many different names that I should shrink from the task, if the duty were assigned me, of collecting all that have been applied to it. But whenever there is much contrariety of opinion among botanists respecting the generic rank and denomination of any plant, I usually resort to its earliest botanical title. Indeed, I feel assured that the nice distinctions upon which later botanists have founded its claims to a different generic position are very much of the same nature as those which divide theologians, whose ecclesiastical acuteness enables them to discern a palpable difference in two doctrinal points, neither of which to an unregenerate mind have any meaning at all. I therefore prefer to call this tree a Mespilus, after Linnæus and Michaux, to save myself the trouble of those infinitesimal investigations that might convince me of the propriety of placing it in every one of a dozen other different genera.

The Shad-bush is a small tree inclining to grow in clumps, instead of making a single stem from the root, and is seldom quite so large or so tall as the mountain ash. The leaves are small and alternate, resembling those of a pear-tree, but more elegant, and covered with a soft silken down on their first appearance; as the foliage ripens, it becomes smooth and glossy. The flowers are white, but without beauty, growing in loose panicles at the ends of the branches. The product of these flowers is a small fruit, about the size of the common wild gooseberry, of a dark crimson color and a very agreeable flavor. This fruit is used very generally in the northern provinces, where the tree is larger and more productive than in New England.

THE CHOKEBERRY.

A SMALLER species of mespilus, familiarly known as the Chokeberry, is more interesting as a flowering plant. It is a slender shrub, with beautiful finely toothed leaves, bearing flowers in clusters very much like those of the hawthorn, with white petals and purple or crimson anthers. The flowers stand erect, but the berries, which are very astringent and are often gathered carelessly with whortleberries, hang from the branches in full pendent clusters. The flowers of this plant are very conspicuous in the latter part of May in all our meadows.

THE MOUNTAIN ASH.

THE Mountain Ash, or Rowan-tree, is beautiful in all its conditions and at all seasons. Its elegant pinnate foliage, not flowing, like that of the locust, but neat, firm, and finely serrate, and its flowers, in large clusters, like those of the elder, render the tree very conspicuous when in blossom. But its greatest ornament is the scarlet fruit that hangs from every branch in the autumn. We could hardly be persuaded to introduce the Mountain Ash into a picture. The primness of its form injures it as a picturesque object in landscape. Its beauty is such as children admire, who are guided by a sense of its material attractions, and do not generally prize a tree except for its elegance and colors. The beauty, however, which attracts the sensual eye in this case is deceitful, for its fruit is of a bitter, sour flavor, and incapable of improvement. European writers say that thrushes are very fond of this fruit. In our land it remains untouched, at least until late in the season, after the black cherries are gone, which tempt all kinds of birds by their superior flavor. The American Mountain Ash differs from the European tree only by its smaller fruit.

I have said that the Mountain Ash is wanting in picturesque qualities; but my remark applies only to its form and habit of growth. On the other hand, it is peculiarly the tree of romance, being remarkable for the many superstitious customs connected with it. According to Evelyn, "There is no churchyard in Wales without a Mountain Ash-tree planted in it, as the yew-trees are in the churchyards of England. So on a certain day of the year everybody in Wales religiously wears a cross made of the wood." Gilpin says that in his time "a stump of the Mountain Ash was generally found in some old burial-place, or near the circle of a Druid's temple, the rites of which were formerly performed under its shade."

Many of the inhabitants of Great Britain still believe that a branch of the Rowan-tree carried about with them is a charm against the evil influences of witchcraft. It is remarkable that similar superstitions connected with this tree prevail among the North American Indians; and it is not improbable that they were introduced by the early Welsh colonists, before the discovery of America by Columbus.

RELATIONS OF TREES TO WATER.

THERE is a spot which I used to visit some years ago, that seemed to me one of the most enchanting of natural scenes. It was a level plain of about ten acres, surrounded by a narrow stream that was fed by a steep ridge forming a sort of amphitheatre round more than half its circumference. The ridge was a declivity of near a hundred feet in height, and so steep that you could climb it only by taking hold of the trees and bushes that covered it. The whole surface consisted of a thin stratum of soil deposited upon a slaty rock; but the growth of trees upon this slope was beautiful and immense, and the water that was constantly trickling from a thousand fountains kept the ground all the year green with mosses and ferns, and gay with many varieties of flowers. The soil was so rich in the meadow enclosed by this ridge, and annually fertilized by the *débris* washed from the hills, that the proprietor every summer filled his barns with hay, which was obtained from it without any cultivation.

I revisited this spot a few years since, after a long period of absence. A new owner, "a man of progress and enterprise," had felled the trees that grew so beautifully on the steep sides of this elevation, and valley and hill have become a dreary and unprofitable waste. The thin soil that sustained the forest, no longer protected by the trees and their undergrowth, has been washed down into the valley, leaving nothing but a bald, rocky surface, whose hideousness is scarcely relieved by a few straggling vines. The valley is also ruined; for the inundations to which it is subject after any copious rain destroy every crop that is planted upon it, and render it impracticable for tillage. It is covered with sand heaps; the little stream that glided round it, fringed with azaleas and wild roses, has disappeared, and the land is reduced to a barren pasture.

The general practice of the pioneers of civilization on this continent was to cut down the wood chiefly from the uplands and the lower slopes of the hills and mountains. They cleared those tracts which were most valuable for immediate use and cultivation. Necessity led them to pursue the very course required by the laws of nature for improving the soil and climate. The first clearings were made chiefly for purposes of agriculture; and as every farm was surrounded by a rampart of woods, it was sheltered from the force of the winds and pleasantly open to the sun. But when men began to fell the woods to supply the demands of towns and cities for fuel and lumber, these clearings were gradually deprived of their shelter, by levelling the surrounding forest and opening the country to the winds from every quarter. But the clearing of the wood from the plains, while it has rendered the climate more unstable, has not been the cause of inundations or the diminution of streams. This evil has been produced by clearing the mountains and lesser elevations having steep or rocky sides; and if this destructive work is not checked by legislation or by the wisdom of the people, plains and valleys now green and fertile will become profitless for tillage or pasture, and the advantages we shall have sacrificed will be irretrievable in the lifetime of a single generation. The same indiscriminate felling of woods has rendered many a once fertile region in Europe barren and uninhabitable, equally among the cold mountains of Norway and the sunny plains of Brittany.

Our climate suffers more than formerly from summer

droughts. Many ancient streams have entirely disappeared, and a still greater number are dry in summer. Boussingault mentions a fact that clearly illustrates the condition to which we may be exposed in thousands of locations on this continent. In the island of Ascension there was a beautiful spring, situated at the foot of a mountain which was covered with wood. By degrees the spring became less copious, and at length failed. While its waters were annually diminishing in bulk, the mountain had been gradually cleared of its forest. The disappearance of the spring was attributed to the clearing. The mountain was again planted, and as the new growth of wood increased, the spring reappeared, and finally attained its original fulness. More to be dreaded than drought, and produced by the same cause, — the clearing of steep declivities of their wood, — are the excessive inundations to which all parts of the country are subject.

If it were in the power of man to dispose his woods and tillage in the most advantageous manner, he might not only produce an important amelioration of the general climate, but he might diminish the frequency and severity both of droughts and inundations, and preserve the general fulness of streams. If every man were to pursue that course which would protect his own grounds from these evils, it would be sufficient to bring about this beneficent result. If each owner of land would keep all his hills and declivities, and all slopes that contain only a thin deposit of soil or a quarry, covered with forest, he would lessen his local inundations from vernal thaws and summer rains. Such a covering of wood tends to equalize the moisture that is distributed over the land, causing it, when showered upon the hills, to be retained by the mechanical action of the trees and their undergrowth of shrubs and herbaceous plants, and by the spongy surface of the soil underneath them, made porous by mosses, decayed leaves, and other *débris*, so that the plains and valleys have a moderate oozing supply of moisture for a long time after every shower. Without this covering, the water when precipitated upon the slopes, would immediately rush down over an unprotected surface in torrents upon the space below.

Every one has witnessed the effects of clearing the woods and other vegetation from moderate declivities in his own neighborhood. He has observed how rapidly a valley is inundated by heavy showers, if the rising grounds that form its basin are bare of trees and planted with the farmer's crops. Even grass alone serves to check the rapidity with which the water finds its way to the bottom of the slope. Let it be covered with bushes and vines, and the water flows with a speed still more diminished. Let this shrubbery grow into a forest, and the valley would never be inundated except by a long-continued and flooding rain. Woods and their undergrowth are indeed the only barriers against frequent and sudden inundations, and the only means in the economy of nature for preserving an equal fulness of streams during all seasons of the year.

At first thought, it may seem strange that the clearing of forests should be equally the cause both of drought and inundations; but these apparently incompatible facts are easily explained by considering the different effects produced by woods standing in different situations. An excess of moisture in the valleys comes from the drainage of the hills, and the same conditions that will cause them to be dried up at certain times will cause them to be flooded at others. Nature's design seems to be to preserve a constant moderate fulness of streams and standing water. This purpose she accomplishes by clothing the general surface of the country with wood. When man disturbs this arrangement, he may produce evil consequences which he had never anticipated. We are not, however, to conclude that we may not improve the soil and climate by changing the original condition of this wooded surface. The clearing of the forest may be reduced to a science whose laws are as sure and unexceptionable as those of mechanics and hydraulics. Though it has not gained much attention from the public mind, it is well understood by the learned who have made this branch of vegetable meteorology their special study. Our danger lies in neglecting to apply these laws to operations in the forest, and in preferring to obtain certain immediate commercial advantages, at the risk of inflicting evils of incalculable extent upon a coming generation.

THE LINDEN-TREE.

THE Lime or Linden tree is generally known among our countrymen as the Bass, and was not, before the present century, employed as a wayside tree. The old standards seen in our ancient villages are European Limes. During the past thirty years the American tree has been very generally planted by roadsides, in avenues and pleasure-grounds, and few trees are more highly valued in these situations. But the American has less beauty than the European tree, which is clothed with softer foliage, has a smaller leaf, and a neater and more elegant spray. Our native Lime bears larger and more conspicuous flowers, in heavier clusters, but of inferior sweetness. Both species are remarkable for their size and longevity. The Lime in Great Britain is a tree of first magnitude, frequently rising to the height of eighty or ninety feet, with a trunk of proportional diameter. The American species is not inferior to it in size or altitude. Some of the largest trees in Western New York are Limes.

The Lime has in all ages been celebrated for the fragrance of its flowers and the excellence of the honey made from them. The famous Mount Hybla was covered with Lime-trees. The aroma from its flowers is like that of mignonette; it perfumes the whole atmosphere, though never disagreeable from excess, and is perceptible to the inhabitants of all the beehives within the circuit of a mile. The Lime is also remarkable for a general beauty of proportion, a bright verdure contrasting finely with the dark-colored branches, and an outline regular and symmetrical without formality. When covered with leaves, it bears some resemblance in outward form to the maple, but surpasses it, when leafless, in the beauty of its ramification. The leaves are roundish heart-shaped, of a clear and lively green in summer, but acquiring a spotted and rusty look in autumn, and adding nothing to the splendors of that season. In the spring, however, no tree of our forest displays a more beautiful verdure before it acquires the uniform dark green of the summer woods.

The branches of the Lime have a very dark-colored surface, distinguishing it from other trees that agree with it in size and general appearance. The bark of the maple, for example, is light and of an ashen-gray tint, and that of the poplars a sort of greenish clay-color. This dark hue renders the spray of the Lime very conspicuous, after a shower, and in spring, when all the leaves are of a light and brilliant green; but these incidental beauties are not very lasting. The branches, being alternate, are very minutely subdivided, and their extremities neatly drawn inwards, so that in a denuded state it is one of our finest winter ornaments. The spray of the beech is more airy, that of the elm more flowing, and that of the oak more curiously netted and interwoven; but the spray of the Lime is remarkable for its freedom from all defect.

George Barnard, who, being a painter, looks upon trees as they are more or less adapted to his own art, remarks —

"When young, or indeed up to an age perhaps of sixty or seventy years, the Lime has a formal appearance, with little variation in its masses of foliage; but let some accident occur, such as the breaking down of a large branch, or the removal of a neighboring tree, it then presents a charming picture."

One of the curiosities of the Lime-tree that deserves notice is a certain winged appendage to the seed, which is a round nut about the size of a pea. This is attached to a long stem, from the end of which, joined to it obliquely, descends a ribbon-like bract, causing it, when it falls, to spin round and travel a long distance upon the wind. If the tree stands on the borders of a pond, where the seeds fall upon the surface, this winged appendage performs the part of a sail, and causes the seeds to be wafted to different points of the opposite shore.

OLD ORCHARDS.

SAUNTERING from the town into solitary field-paths, and passing by rustic cottages with their pleasant array of haystacks, unornamented barns, and simple gardens full of roses and sunflowers, and wending our way over pebbly hills and plashy hollows, we enter a recent growth of wood that has come up spontaneously upon an old neglected farm. We follow a wood-path, shaded by a stunted growth of pines and white birches, and bordered with wild-flowers, and, leaving the ruins of an old cider-mill, reach an opening, enclosed by a dilapidated stone-wall, half concealed by tall shrubs and vines and by trees that have encroached upon its boundaries. Emerging into this open space we find ourselves in an old orchard that still bears meagre crops of fruit, which was an appendage to a farm long neglected and abandoned, now half restored to its original condition as a forest.

I have often called the attention of lovers of nature to the peculiar beauty which is apparent in an old orchard. I know it is not much admired by improvers. It has neither trimness nor elegance. There is nothing in the style of the trees or the character of the ground that awakens any ideas of aristocratic dignity. The old stone-walls that enclose it, loose and dilapidated, betray no extravagant outlay of money. They remind you only of the simple labor of hard hands and the rude husbandry of toiling men. The ideas associated with the old orchard are those of rustic simplicity, — of apple-gathering by rural swains; of golden, russet, and crimson fruit,

hanging from the trees or lying in heaps upon the ground; of juvenile sports; of truancy and birdnesting; of the cider-mill and the apple-cart, and of hard bargaining by men whose accumulations can be made only by penurious saving.

The chief beauty of the apple-tree — that kind of beauty which affects the eye, rather than the mind — comes from its flowers and fruit. But when it has grown hoary with the mosses of age, there is no small tree that displays so many picturesque qualities, in which it is hardly surpassed by the oak. The apple-tree, indeed, has many of the characters of the oak. Though of inferior size, it displays great breadth in proportion to its height, and a corresponding sturdiness of appearance, caused by the length and horizontal spread of its lower branches. There is an air of picturesque antiquity about an old orchard which we look for in vain among the trees of our forest. Whether this comes from the greater age of the fruit-trees, which have been preserved through many generations, while the neighboring forest growth is young and new, or from the sturdy and rugged forms of the apple-trees, I will confess that I look to the old orchard, rather than to the wood, to gratify my taste for the picturesque in trees.

One of the most charming appearances about an old orchard is neglect. I cannot account for all the pleasure we feel, under certain circumstances, from the sight of a neglected field or old family estate; but I have often felt it to be the cause of some of the most agreeable sensations that can be awakened by the scenes of nature. Half the pleasure that comes from the sight of moss has this metaphysical basis. Moss is one of the evidences of antiquity, and yields to a natural object a venerable look, as gray hairs add dignity to the head of an old man. It is also one of the tokens of freedom and seclusion. We know, when we see it, there is no watchful guardian of the place, anxious to keep out trespassers, from the bird that sings in the tree, to the truant boy who watches for forbidden fruit. It is a sign that the field has been partially restored to nature; that it is in one sense abandoned by its proprietor, and is freely open to all, like the wild wood and the whortleberry pasture.

The grounds have acquired a rugged and bushy appearance from the renewal of many wild plants. Viburnums, elders, and wild roses have reappeared in the borders, and tufts of low blueberry-bushes are conspicuous here and there, fringing the edges of some projecting rock with their heathlike blossoms and foliage and their bright azure fruit. The clover and herdsgrass have disappeared, and in the place of them are brown spleenworts and tufted andropogons. The rocks forming the loose stone-wall are incrusted with lichens and garlanded with the creeping sumach, and the squirrel finds shelter and concealment there as peaceful as in his native wild wood. We find here some of the solitary birds of the forest, and the quail leads out her young fearlessly, as if the old apple-trees were guardians of their safety. We look around, doubtful if human hands have ever marred the sacredness of this lovely spot; but we see that it is a quadrangular space, that the old trees stand in broken rows; and the sweet pyrola and the red summer lily announce by their presence the restoration of the grounds to nature.

Apple-trees are not the only denizens of an old orchard. Around its borders standard forest trees fling their shadowy branches over the old stone-wall; equalling perhaps the fruit-trees in age, though from their greater vitality still retaining the vigor and beauty of their prime. The oak is there, with its playful squirrels and its contorted boughs. A few hickories and wild-cherry trees stand there, like sentinels of the field; and now and then a white-pine rears its summit above all the other trees, exceeding even the oak in grandeur. All these forest trees, with a variety of undershrubs, enclose the old orchard, hide it from distant view, and render it a sacred precinct where we may saunter as in an old graveyard, and read from the parasitic mosses and lichens, and the hieroglyphical patches and incrustations, many a quaint incident in the history of the old trees.

While strolling among these old trees, we are struck by their expressions of dignity and pathos. There is about them also a rusticity not to be discovered in a young and thriving orchard. Upon the moss-covered bark of the aged apple-trees, and in the hollow branches, that afford a retreat to the woodpecker and the bluebird, we see a picture of that rudeness we observe in the rocks that project from the sides of the steep hills and crown their summits. We feel easy in their company, as in the presence of the old yeoman who has always garnered their fruit. The operations of tillage have been almost obliterated by crowds of spontaneous wildings. We look upon the wide and frequent clumps of moss upon the greensward, interspersed with low shrubs and groups of wild-flowers, and we say, "Here we may repose, under the shade of these trees, and hear the voice of the Echo as before she was banished to her shell, and breathe the incense burned by the Dryad in her own temple."

In early summer the singing-birds carol their rapid notes in multitudes, crossing and recrossing the field and careering in circles over the trees, and pouring out their wild notes while on the wing in the joy of that freedom designed by Nature for all her creatures. An old orchard is the favorite resort of birds at all seasons. More nests are built in these rugged apple-trees than in any other grove of equal size, especially if there be woods and thickets around to afford them shelter and seclusion. The density of the spray of the apple-trees, and the many irregularities of their ramification, create little hollows and angles for the nests of birds that build in the open air, and their hollow boughs form safe retreats for those species that seek protection as it were under a roof.

THE KALMIA.

THE Kalmia, on account of its superficial resemblance to the green bay-tree, often called the American laurel, is more nearly allied to the heath. The name of Kalmia, which is more musical than many others of similar derivation, was given to this genus of evergreen shrubs by Linnæus, in honor of Peter Kalm, a distinguished botanist and one of his pupils. This is exclusively an American family of plants, containing only five species, three of which are natives of New England soil and two of them among our most common shrubs.

THE MOUNTAIN LAUREL.

Not one of our native shrubs is so generally admired as the Mountain Laurel; no other equals it in glowing and magnificent beauty. But the "patriots" who plunder the fields of its branches and flowers for gracing the festivities of the "glorious Fourth" will soon exterminate this noble plant from our land. There are persons who never behold a beautiful object, especially if it be a flower or a bird, without wishing to destroy it for some selfish, devout, or patriotic purpose. The Mountain Laurel is not so showy as the rhododendron, with its deeper crimson bloom; but nothing can exceed the minute beauty of its individual flowers, the neatness of their structure, and the delicacy of their shades as they pass from rose-color to white on different bushes in the same group. The flower is monopetalous, expanded to a cup with ten angles and scalloped edges. "At the circumference of the disk on the inside," says Darwin, "are ten depressions or pits, accompanied with corresponding prominences on the outside. In these depressions the anthers are found lodged at the time when the flower expands. The stamens grow from the base of the corolla, and bend outwardly, so as to lodge the anthers in the cells of the corolla. From this confinement they liberate themselves, during the period of flowering, and strike against the sides of the stigma." This curious internal arrangement of parts renders the flower very beautiful on close examination. The flowers are arranged in flat circular clusters at the terminations of the branches.

We seldom meet anything in the forest more attractive than the groups of Mountain Laurel, which often cover extensive slopes, generally appearing on the edge of a wood, and becoming more scarce as they extend into the interior or wander outwardly from the border. But if we meet with an opening in the wood where the soil is favorable, — some little sunny dell or declivity, — another still more beautiful group opens on the sight, sometimes occupying the whole space. The Mountain Laurel does not constitute the undergrowth of any family of trees, but avails itself of the protection of a wood where it can flourish without being overshadowed by it. In the groups on the outside of the wood, the flowers are usually of a fine rose-color, fading as they are more shaded, until in the deep forest we find them, and the buds likewise, of a pure white. I am not acquainted with another plant that is so sensitive to the action of light upon the color of its flowers. The buds, except in the dark shade, before they expand, are of a deeper red than the flowers, and hardly less beautiful.

The Mountain Laurel delights in wet places, in springy lands on rocky declivities where there is an accumulation of soil, and in openings surrounded by woods, where the land is not a bog, but wet enough to abound in ferns. In such places the Kalmia, with its bright evergreen leaves, forms elegant masses of shrubbery, even when it is not in flower. Indeed, its foliage is hardly less conspicuous than its flowers. I believe the Kalmias are not susceptible of modification by the arts of the florist. Nature has endowed them with a perfection that cannot be improved.

THE LOW LAUREL, OR LAMBKILL.

The low Laurel, or small Kalmia, is plainly one of nature's favorite productions; for, the wilder and ruder the situation, the more luxuriant is this plant and the more beautiful are its flowers. These are of a deep rose-color, arranged in crowded whorls around the extremities of the branches, with the recent shoot containing a tuft of newly formed leaves surmounting each cluster of flowers. This plant, through not celebrated in horticultural literature or song, is one of the most exquisite productions of nature. Many other shrubs which are more showy are not to be compared with this in the delicate structure of its flowers and in the beauty of their arrangement and colors. Of this species the most beautiful individuals are found on the outer edge of their groups.

There has been much speculation about the supposed poisonous qualities of this plant and its allied species. Nuttall thought its flowers the source of the deleterious honey discovered in the nests of certain wild bees. There is also a general belief that its leaves are poisonous to cattle and flocks. But all positive evidence is wanting to support any of these notions. The idea associated with the name of this species is a vulgar error arising from a corruption of the generic name, from which Lambkill may be thus derived, — Kalmia, Kallamia, Killamia, Killam, Lambkill. There is no other way of explaining the origin of its common English name. I have never been able to discover an authentic account, and have never known an instance of the death of a sheep or a lamb from eating the leaves of this plant. It is an error having its origin in a false etymology; and half the notions that prevail in the world with regard to the medical virtues and other properties of plants have a similar foundation.

It is stated in an English manual of Medical Botany that the brown powder that adheres to the petioles of the different species of Kalmia, Andromeda, and Rhododendron is used by the North American Indians as snuff.

MOTIONS OF TREES.

WHILE Nature, in the forms of trees, in the color of their foliage and the gracefulness of their spray, has displayed a great variety of outline and tinting, and provided a constant entertainment for the sight, she has increased their attractions by endowing them with a different susceptibility to motion from the action of the winds. In their motions we perceive no less variety than in their forms. The different species differ like animals; some being graceful and easy, others stiff and awkward; some calm and intrepid, others nervous and easily agitated. Perhaps with stricter analogy we might compare them to human beings; for we find trees that represent the man of quiet and dignified deportment, also the man of excited manners and rapid gesticulations. Some trees, like the fir, having stiff branches and foliage, move awkwardly backward and forward in the wind, without any separate motions of their leaves. While we admire the symmetrical and stately forms of such trees, we are reminded of men who present a noble personal appearance, accompanied with ungainly manners.

Some trees, having stiff branches with flexible leaves, do not bend to a moderate breeze, but their foliage readily yields to the motion of the wind. This habit is observed in the oak and the ash, in all trees that have a pendulous foliage and upright or horizontal branches. The poplars possess this habit in a remarkable degree, and it is proverbial in the aspen. It is also conspicuous in the common pear-tree and in the small white-birch. Other trees, like the American elm, wave their branches gracefully, with but little apparent motion of their leaves. We observe the same habit in the weeping willow, and indeed in all trees with a long and flexible spray. The wind produces by its action on these a general sweeping movement without any rustle. It is easy to observe, when walking in a grove, that the only graceful motions come from trees with drooping branches, because these alone are long and slender.

The very rapid motion of the leaves of the aspen has given origin to some remarkable superstitions. The Highlanders of Scotland believe the wood of this tree to be that of which the holy cross was made, and that its leaves are consequently never allowed to rest. Impressed with the awfulness of the tragedy of the crucifixion, they are constantly indicating to the winds the terrors that agitate them. The small white birch displays considerable of the same motion of the leaves; but we take little notice of it, because they are softer and produce less of a rustling sound. The flickering lights and shadows observed when walking under these trees, on a bright noonday, have always been admired. All these habits awaken our interest in trees and other plants by assimilating them to animated things.

Much of the beauty of the silver poplar comes from its glittering lights, when it presents the green upper surface of its foliage, alternating rapidly with the white silvery surface beneath. This we may readily perceive even in cloudy weather, but in the bright sunshine the contrasts are very brilliant. In all trees, however, we observe this glittering beauty of motion in the sunshine. The under part of leaves being less glossy than the upper part, there is in the assemblage the same tremulous lustre that appears on the rippled surface of a lake by moonlight.

We observe occasionally other motions which I have not described, such as the uniform bending of the whole tree. In a strong current of wind, tall and slender trees especially attract our attention by bending over uniformly like a plume. This habit is often seen in the white birch, a tree that in its usual assemblages takes a plume-like form. When a whole grove of white birches is seen thus bending over in one direction from the action of a brisk wind, they seem like a procession of living forms. In a storm we watch with peculiar interest the bending forms of certain tall elms, such as we often see in clearings, with their heads bowed down almost to the ground by the force of the tempest. It is only the waves of the ocean and the tossing of its billows that can afford us so vivid an impression of the sublimity of a tempest as the violent swaying of a forest and the roaring of the winds among the lofty tree-tops.

The motions of an assemblage of trees cannot be observed except from a stand that permits us to look down upon the surface formed by their summits. We should then perceive that pines and firs, with all the stiffness of their branches, display a great deal of undulating motion. These undulations or wavy movements are particularly graceful in a grove of hemlocks, when they are densely assembled without being crowded. It is remarkable that one of the most graceful of trees belongs to a family which are distinguished by their stiffness and formality. The hemlock, unlike other firs and spruces, has a very flexible spray, with leaves also slightly movable, which are constantly sparkling when agitated by the wind. If we look down from an opposite point, considerably elevated, upon a grove of hemlocks when they are exposed to brisk currents of wind, they display a peculiar undulating movement of the branches and foliage, made more apparent by the glitter of their leaves.

The surface of any assemblage of trees when in motion bears a close resemblance to the waves of the sea. But hemlocks represent its undulations when greatly agitated, without any broken lines upon its surface. Other firs display in their motions harsher angles and a somewhat broken surface of the waves. We see the tops of these trees and their extreme branches awkwardly swaying backwards and forwards, and forming a surface like that of the sea when it is broken by tumultuous waves of a moderate height. The one suggests the idea of tumult and contention; the other, that of life and motion combined with serenity and peace.

THE TULIP-TREE.

THE Tulip-tree is pronounced by Dr. Bigelow "one of the noblest trees, both in size and beauty, of the American forest." It certainly displays the character of immensity,—a quality not necessarily allied with those features we most admire in landscape. It is not very unlike the Canada poplar, and is designated by the name of White Poplar in the Western States. The foliage of this tree has been greatly extolled, but it has the heaviness which is apparent in the foliage of the large-leaved poplars, without its tremulous habit. The leaves, somewhat palmate in their shape, are divided into four pointed lobes, the middle rib ending abruptly, as if the fifth lobe had been cut off. The flowers, which are beautiful, but not showy, are striped with green, yellow, and orange. They do not resemble tulips, however, so much as the flowers of the abutilon and althea.

This tree is known in New England rather as an ornamental tree than as a denizen of the forest. Its native habitats are nearly the same with those of the magnolia, belonging to an allied family. There is not much in the proportions of this tree to attract our admiration, except its size. But its leaves are glossy and of a fine dark green, its branches smooth, and its form symmetrical. It is a tree that agrees very well with dressed grounds, and its general appearance harmonizes with the insipidity of artificial landscape. It is wanting in the picturesque characters of the oak and the tupelo, and inferior in this respect to the common trees of our forest.

THE MAGNOLIA.

THE Magnolia, though, excepting one species, a stranger to New England soil, demands some notice. Any one who has never seen the trees of this genus in their native soil can form no correct idea of them. I would not say, however, that they would fall short of his conceptions of their splendor. When I first beheld one of the large magnolias, though it answered to my previous ideas of its magnificence, I thought it a less beautiful tree than the Southern cypress, and less picturesque than the live-oak, the black-walnut, and some other trees I saw there. The foliage of the Magnolia is very large and heavy, and so dark as to look gloomy. It is altogether too sombre a tree in the open landscape, and must add to the gloom of any wood which it occupies, without yielding to it any other striking character.

There are several species of Magnolia cultivated in pleasure-grounds, the selection being made from those bearing a profusion of flowers. The only one that grows wild in New England is of small stature, sometimes called the Beaver-tree. It inhabits a swamp near Gloucester, about twenty miles from Boston. This place is its northern boundary. The flowers are of a dull white, without any beauty, but possessed of a very agreeable fragrance, causing them to be in great demand. The Magnolia wood is annually stripped both of flowers and branches, and the trees will probably be extirpated before many years by this sort of vandalism.

THE PICTURESQUE.

The picturesque is that visible quality in any scene or object which, without positive beauty, awakens an agreeable sentiment; or, in other words, any quality in an ugly or homely scene that yields it the charm of one that is beautiful. The picturesque is, indeed, only a synonyme of *relative beauty*. When we have learned the connection between certain scenes and certain interesting events or incidents, and these have become familiarly associated in our minds, we may be affected more agreeably by the sight of them than by the most beautiful colors and forms. But the word "picturesque" cannot, in strict accordance with its literal meaning, be applied to scenes of indefinite extent. Properly it should be used to designate only limited portions of a landscape, capable of being easily embraced in a picture. But it is very generally applied to either narrow or extensive prospects, to express certain qualities which would be more definitely described as wild, dreary, pastoral, or romantic. Like other abstract terms, applied to the face of nature, it has a vague signification, and conveys a poetical rather than a distinct image to the mind.

Rudeness seems to be a favorable groundwork for the representation of rural objects, but it is not a necessary ingredient of the picturesque. A huge rock in mid-ocean, exposed to the fury of winds and waves, is sufficiently rude, but it would require certain accompaniments to render it poetical or interesting. If there were a lighthouse upon this rock, containing one solitary family, it would partake of this quality in a high degree; and every little appurtenance of trees and shrubbery, of garden and flower-beds, and of domestic animals, would heighten this interest. The perilous character of the situation prepares the mind to sympathize with the inhabitants thus isolated from the world and exposed to the dangers of the sea. The rudeness of the rock increases our interest by suggesting the simple habits of the occupants. Rudeness, indeed, is commonly a heightening of the picturesque, implying the absence of that sort of cultivation which is associated with the extremely unpicturesque objects of fashion and pride.

All great paintings appeal to the affections of the human heart by representing a single scene that powerfully affects the sympathy or the imagination. A solitary traveller seeking protection in a cave or a ruin from some besetting danger, and a group of sailors exposed on a raft in mid-ocean, are sympathetic pictures; while a ruined temple or castle appeals to the imagination and to our reverence for antiquity. A weary pilgrim by the wayside, leaning on his staff, is a scene for the canvas, while a king on his throne is adapted only to the dry page of the historian. But let this king be dethroned, and seek protection in a woodman's hut, and he becomes in that situation one of the most affecting subjects for painting or romance. As soon as the great have fallen into adversity, they are candidates for our sympathy; and we feel the more interest in them in proportion as they have fallen from a lofty eminence. They are now reduced to the common level of fortune, while their former greatness seems to render them worthy of a better fate.

Those scenes in real landscape, as well as in paintings, yield us the most pleasure which are comprised within narrow limits. Panoramas are seldom very interesting. As our vision is extended beyond certain bounds, we approach more nearly to views that awaken a sense of dreariness or grandeur. When ascending a mountain we are commonly affected with the most pleasure by viewing some limited prospect from a small altitude. An opening in a wood reveals a single farm-house, with its outbuildings, its green and yellow fields of tillage, flocks grazing on the opposite slope, and here and there a human being engaged in some pleasant occupation. As we continue our ascent, this farm and cottage soon become an insignificant portion of an almost boundless variety of objects. The attention now becomes unfixed. The mind rests agreeably on no particular scene, but is somewhat exhilarated by the grandeur of the whole prospect. The spectacle is no longer picturesque: it is wild, rude, desolate, or sublime; but there is nothing in it that awakens the sympathetic interest excited by the first limited view of the farm and cottage.

A scene in real nature, to be picturesque, must be suggestive; and in a picture or in a poem we must not be presented with all which the poet or the artist would suggest to the mind. By this law of agreeable effects we may account for the pleasantness of a crooked or winding road and the tiresomeness of a straight road, which, by displaying the approaching scenes to view, leaves nothing to the imagination. The picturesque quality of an object is made up in great part of its suggestiveness of agreeable images not beheld; the "evidence of things not seen," and those of an interesting nature. And if it be a picture, the ideal perfection of the work depends on an ingenious selection of those points which call up the most charming images in the mind.

Something of human interest must be associated with a scene in nature to render it picturesque in the highest degree. A rude scene, without qualifying accompaniments, reminds us only of waste and discomfort. To afford it a sympathetic character, there should be added a little hut, a stranded vessel, a grave, a monument, or some other object allied to humanity. The sea itself, with all its grandeur and sublimity, is not picturesque; but it is rendered such by a ship or a lighthouse. There is nothing of this quality in a naked representation of the polar ices; but add two graves and headstones to the scene, as discovered in a late Arctic Expedition, and the sorrowful human interest thus associated with it renders it picturesque beyond all other objects. The most remarkable scene in nature, if exhibited in a painting, without some domestic animal or human being, a cottage, or some simple work of human hands, would be cold and unaffecting. Thus in poetic descriptions of Nature, she is personified to increase our interest. We personify the sun, the moon, the seasons, the months, and the hours. The personification likewise of abstract ideas, by linking them with humanity, causes them to take a stronger hold on the imagination.

Hence we may account for the introduction of many circumstances into pictures, that seem trifling and insignificant. Of all things in the world, if considered without reference to other objects, smoke is the last we should call picturesque. But as the beauty of all this class of objects is merely relative, smoke, when issuing on a still morning from the chimney of a cottage, renders it more lively and poetical, by indicating that it is inhabited and that its inmates are stirring within, and thereby increasing its expression of humanity. One of Moore's songs owes its popularity to a picturesque allusion to smoke in the first stanza:—

"I knew by the smoke that so gracefully curled
Above the green elms, that a cottage was near;
And I said, 'If there 's peace to be found in the world,
The heart that is humble might hope for it here.'"

Some writers refer the pleasure derived from the sight

of smoke to "the waving line of beauty." This is but the cant of metaphysical pedants, who delight in tracing effects to inadequate causes. As well might we refer the emotions we feel when watching the lightning that flashes through a dark cloud to "the zigzag line of sublimity." The waving line, or any other line of beauty, must be associated with some agreeable sentiment, or it is nothing. There is no positive beauty in smoke. It owes the interest it awakens entirely to association. Even when it ascends spirally from a mere brick-kiln, its beauty comes from its indication of the pleasant serenity of the atmosphere, that saves it from blending into a formless mass, not from the character of the figure it displays.

To a boor nothing is either poetical or picturesque. No object is attractive to him that does not either dazzle his eyes or excite his astonishment. Scenes that would suggest a thousand delightful images to a man of cultivated imagination are to the boor a mere blank. To him a rock, a tree, and a house have no connection with sentiment. If the rock does not reach to the clouds, if the tree does not rear itself stupendously into the air, or if the house is not magnificent in size or embellishments, they are nothing to him, because they excite no admiration. But to a man of feeling a mere hovel awakens the most agreeable fancies, and a sheepfold may cause more interest than a palace. A rock presents him with a whole field of natural curiosities, and a tree becomes an object under whose shade he may revive a long train of studious recreations.

THE LOCUST.

THE waysides in the Middle States do not contain a more beautiful tree than the Locust, with its profusion of pinnate leaves and racemes of flowers that fill the air with the most agreeable odors. In New England the Locust is subject to the ravages of so many different insects that it is commonly stinted in its growth, its branches withered and broken, and its symmetry destroyed. But the deformities produced by the decay of some of its important limbs cannot efface the charm of its fine pensile foliage. In winter it seems devoid of all those proportions we admire in other trees. It rears its tall form, withered, shapeless, and deprived of many valuable parts, without proportional breadth, and wanting in any definite character of outline. Through all the early weeks of spring we might still suppose it would never recover its beauty. But May hangs on those withered boughs a green drapery that hides all their deformity; she infuses into their foliage a perfection of verdure that no other tree can rival, and a beauty in the forms of its leaves that renders it one of the chief ornaments of the groves and waysides. June weaves into this green leafage pendent clusters of flowers of mingled brown and white, filling the air with fragrance, and enticing the bee with odors as sweet as from groves of citron and myrtle.

The finely cut and delicate foliage of the Locust and its jewelled white flowers, hanging gracefully among its dark green leaves, yield it a peculiar style of beauty, and remind us of some of the finer vegetation of the tropics.

The leaflets, varying in number from nine to twenty-five on a common stem, have a singular habit of folding over each other in wet and dull weather and in the night, thus displaying a sensitiveness that is remarkable in all the acacia family. The Locust is not highly prized by landscape gardeners, who cannot reconcile its defects to their serpentine walks and their velvety lawns. But I am not sure that the accidental deformities of the Locust may not contribute to its picturesque attractions, when, for example, from its furrowed and knotted trunk a few imperfect limbs project, and suspend over our heads a little canopy of the finest verdure.

Phillips says of the Locust, that, when planted in shrubberies, it becomes the favorite resort of the nightingale, to avail itself of the protection afforded by its thorns. There are many other small birds that seek the protection of thorny bushes for their nests. On the borders of woods, a barberry or hawthorn bush is more frequently selected by the catbird and the yellow-throat than any other shrub. I have observed that the indigo-bird shows a remarkable attachment to the Locust, attracted, perhaps, by some favorite insect that lives upon it. The only nests of this bird I have ever discovered were in the branches of the Locust. It is worthy of notice, that, notwithstanding its rapid and thrifty growth in favorable situations, this tree never occupies exclusively any large tracts of country. It is found only in small groups, scattered chiefly on the outside of woods containing different species. The foliage of the Locust, like that of other leguminous plants, is very fertilizing to the soil, causing the grassy turf that is shaded by this tree to be always green and luxuriant.

THE HONEY LOCUST.

The Honey Locust is not an uncommon tree in the enclosures of suburban dwellings, and by the waysides in many parts of the country. Some of them have attained a great height, overtopping all our shade-trees except the elm and the oak. This tree in June bears flowers without any beauty, hanging from the branches in small greenish aments. The outer bark peels from the trunk, like that of the shellbark hickory. The thorns investing the trunk as well as the boughs are very singular, consisting of one long spine with two and sometimes more shorter ones projecting out of it, like two little branches, near its base. Three is the prevailing number of these compound thorns. Hence the name of Three-Thorned Acacia applied to the Honey Locust.

This tree bears some resemblance to the common Locust but its leaflets are smaller, and of a lighter green. It is not liable, however, to the attacks of insects, and is seen, therefore, in all its normal and beautiful proportions. It displays much of the elegance of the tropical acacias in the minute division and symmetry of its compound leaves. These are of a light and brilliant green, and lie flat upon the branches, giving them a fan-like appearance, such as we observe in the hemlock. Though its principal branches are given out at an acute angle, many of them are horizontal, extending outwards with frequent contortions. The Honey Locust derives its name from the sweetness of the pulp that envelops the seeds contained in their large flat pods. This tree is not an uncommon hedge plant in Massachusetts, but it is not found in the New England forest. Its native region is the wide valley between the Alleghanies and the Mississippi River.

RELATIONS OF TREES TO THE ATMOSPHERE.

I HAVE not much faith in the science of ignorant men; for the foundations of all knowledge are laid in books; and those only who have read and studied much can possess any considerable store of wisdom. But there are philosophers among laboring swains, whose quaint observations and solutions of nature's problems are sometimes worthy of record. With these men of untutored genius I have had considerable intercourse, and hence I oftener quote them than the learned and distinguished, whom I have rarely met. The ignorant, from want of knowledge, are always theorists; but genius affords its possessor, how small soever his acquisitions, some glimpses of truth which may be entirely hidden from the mere pedant in science. My philosophic friend, a man of genius born to the plough, entertained a theory in regard to the atmosphere, which, though not strictly philosophical, is so ingenious and suggestive that I have thought an account of it a good introduction to this essay.

My friend, when explaining his views, alluded to the well-known fact that plants growing in an aquarium keep the water supplied with atmospheric air — not with simple oxygen, but with oxygen chemically combined with nitrogen — by some vital process that takes place in the leaves of plants. As the lungs of animals decompose the air which they inspire, and breathe out carbonic-acid gas, plants in their turn decompose this deleterious gas, and breathe out pure atmospheric air. His theory is that the atmosphere is entirely the product of vegetation, and that nature has no other means of composing it; that it is not simply a chemical, but a vital product; and that its production, like its preservation, depends entirely on plants, and would be impossible without their agency. But as all plants united are not equal in bulk to the trees, it may be truly averred that any series of operations or accidents that should deprive the earth entirely of its forests would leave the atmosphere without a source for its regeneration.

The use of the foliage of trees in renovating the atmosphere is not, I believe, denied by any man of science. This theory has been proved to be true by experiments in vital chemistry. The same chemical appropriation of gases and transpiration of oxygen is performed by all classes of vegetables; but any work in the economy of nature assigned to vegetation is the most effectually accomplished by trees. The property of foliage that requires carbonic-acid gas for its breathing purposes, and causes it to give out oxygen, is of vital importance; and it is hardly to be doubted that a close room well lighted by the sun would sustain its healthful atmosphere a longer time, if it were filled with plants in leaf, but not in flower, and occupied by breathing animals, than if the animals occupied it without the plants.

But there is another function performed by the foliage of trees and herbs in which no chemical process is involved, — that of exhaling moisture into the atmosphere after it has been absorbed by the roots. Hence the humidity of this element is greatly dependent on foliage. A few simple experiments will show how much more rapidly and abundantly this evaporation takes place when the soil is covered with growing plants than when the surface is bare. Take two teacups of equal size and fill them with water. Place them on a table, and insert into one of them cuttings of growing plants with their leaves, and let the other stand with water only. In a few hours the water will disappear from the cup containing the plants, while that in the other cup will not be sensibly diminished. Indeed, there is reason to believe that gallons of water might be evaporated into the air by keeping the cup containing the cuttings always full, before the single gill contained in the other cup would disappear. If a few cuttings will evaporate a half-pint of water in twelve hours, we can imagine the vast quantity constantly exhaled into the atmosphere by a single tree. The largest steam-boiler in use, kept constantly boiling, would not probably evaporate more water than one large elm in the same time.

We may judge, from our experiment with the cuttings, that a vastly greater proportion of moisture would be exhaled into the atmosphere from any given surface of ground when covered with vegetation, than from the same amount of uncovered surface, or even of standing water. Plants are indeed the most important existing agents of nature for conveying the moisture of the earth into the air. The quantity of transpiring foliage from a dense assemblage of trees must be immense. The evaporation of water from the vast ocean itself is probably small compared with that from the land which it surrounds. And there is reason to believe that the water evaporated from the ocean would not produce rain enough to sustain vegetation, if by any accident every continent and island were deprived of its trees. The whole earth would soon become a desert. I would remark, in this place, that trees are the agents by which the superfluous waters of the ocean, as they are supplied by rivers emptying into it, are restored to the atmosphere and thence again to the surface of the earth. Trees pump up from great depths the waters as they ooze into the soil from millions of subterranean ducts ramifying in all directions from the bed of the ocean.

We see a double system of operations carried on invisibly by trees and other plants. By them the moisture of the earth is distilled into the air, where it is converted into clouds, and returned to the earth again in the form of dews and rain. Every fall of rain purifies the air through which it descends, and carries down with it into the soil a fresh supply of azote and oxygenated air, which is needful to the roots of the plants. This constant exhalation and absorption resembles the breathing of animals, except that with plants inspiration and expiration are simultaneous, and not alternate. I would venture to say that when the sciences of vital chemistry and vegetable meteorology are carried to perfection, the fruitfulness of any region below the Arctic Circle may be established and preserved by the systematic operations of man upon the forest.

In all temperate latitudes where man has not counteracted the efforts of Nature by his own works, she has covered the land with forest as the most effectual means of preserving a constant circulation of moisture between the earth, the atmosphere, and the ocean. But this primitive condition of the earth is not the one most favorable to the wants of civilized man. There is a certain relative disposition, as well as proportion of wood, pasture, and tillage, that would improve the climate for man's purposes, and another that would injure it. Nature clothes all parts with trees, and leaves it to man to improve or to ruin the climate, according as he is wise or stupid. Nations in most cases have ruined it, and then sunk into barbarism; for civilization has never, in any country, long survived the destruction of its forests.

THE HOLLY.

As the hawthorn is consecrated to vernal festivities, the Holly is sacred to those of winter, and the yew to those attending the burial of the dead. In Europe, from the earliest ages, the Holly has been employed for the decoration of churches during Christmas. The poets have made it a symbol of forethought, because its leaves are saved from the browsing of animals by the thorns that surround them; and the berries, concealed by its prickly foliage, are preserved for the use of the winter birds. The Holly is found only in the southern parts of New England. In Connecticut it is common, and in the Middle and Southern States it is a tree of third magnitude. The leaves of the Holly are slightly sinuate or scalloped, and furnished at each point with short spines. It not only retains its foliage in the winter, but it loses none of that brilliancy of verdure that distinguishes it at other seasons.

There seems to be no very notable difference between the American and European Holly. Selby says of the latter: "The size which the Holly frequently attains in a state of nature, as well as when under cultivation, its beauty and importance in forest and woodland scenery, either as a secondary tree or merely as an underwood shrub, justify our placing it among the British forest trees of the second rank." He adds: "As an ornamental evergreen, whether in the form of a tree or as an undergrowth, the Holly is one of the most beautiful we possess. The deep green glittering foliage contrasts admirably with the rich coral hue of its berries."

THE SPIRÆA.

In the month of July the wooded pastures are variegated with little groups of shrubbery full of delicate white blossoms in compound pyramidal clusters, attracting more attention from a certain downy softness in their appearance than from their beauty. These plants have received the name of Spiræa from the spiry arrangement of their flowers. The larger species among our wild plants, commonly known as the Meadow-Sweet, in some places as Bridewort, is very frequent on little tussocks and elevations rising out of wet soil. It is a slender branching shrub, bearing a profusion of small, finely serrate and elegant leaves, extending down almost to the roots, and a compound panicle of white impurpled flowers at the ends of the branches. It is well known to to all who are familiar with the wood-scenery of New England, and is seen growing abundantly in whortleberry pastures, in company with the small kalmia and the swamp rose. It is a very free bloomer, lasting from June till September, often blending a few solitary spikes of delicate flowers with the tinted foliage of autumn.

THE HARDHACK.

The flowers of the purple Spiræa, or Hardhack, are conspicuous by roadsides, especially where they pass over wet grounds. It delights in the borders of rustic wood-paths, in lanes that conduct from the enclosures of some farm cottage to the pasture, growing all along under the loose stone-wall, where its crimson spikes may be seen waving in the wind with the nodding plumes of the golden-rod and the blue spikes of the vervain, well known as the "Simpler's Joy." The Hardhack affords no less pleasure to the simpler, who has used its flowers from immemorial time as an astringent anodyne. There is no beauty in any part of this plant, except its pale crimson flowers, which are always partially faded at the extremity or unopened at the base, so that a perfect cluster cannot be found. The leaves are of a pale imperfect green on the upper surface and almost white beneath, and without any beauty. The uprightness of this plant, and the spiry form of its floral clusters, has gained it the name of "Steeplebush," from our church-going ancestors.

THE HAWTHORN.

Few trees have received a greater tribute of praise from poets and poetical writers than the Hawthorn, which in England especially is consecrated to the pastoral muse and to all lovers of rural life. The Hawthorn is also a tree of classical celebrity. Its flowers and branches were used by the ancient Greeks at wedding festivities, and laid upon the altar of Hymen in the floral games of May, with which from the earliest times it has been associated. In England it is almost as celebrated as the rose, and constitutes the most admired hedge-plant of that country. It is, indeed, the beauty of this shrub that forms the chief attraction of the English hedge-rows, which are not generally clipped, but allowed to run up and bear flowers. These are the principal beauties of the plant; for its leaves are neither luxuriant nor flowing.

The Hawthorn in this country is not associated with hedge-rows, which with us are only matters of pride and fancy, not of necessity, and their formal clipping causes them to resemble nature only as a wooden post resembles a tree. Our admiration of the Hawthorn, therefore, comes from a pleasant tradition derived from England, through the literature of that country, where it is known by the name of May-bush, from its connection with the floral festivities of May. The May-pole of the south of England is always garlanded with its flowers, as crosses are with holly at Christmas. The Hawthorn is well known in this country, though unassociated with any of our rural customs. Many of its species are indigenous in America, and surpass those of Europe in the beauty of their flowers and fruit. They are considered the most ornamental of the small trees in English gardens.

The flowers of the Hawthorn are mostly white, varying in different species through all the shades of pink, from a delicate blush-color to a pale crimson. The fruit varies from yellow to scarlet. The leaves are slightly cleft, like those of the oak and the holly. The flowers are produced in great abundance, and emit an agreeable odor, which is supposed by the peasants of Europe to be an antidote to poison.

SUMMER WOOD-SCENERY.

I HAVE alluded to a beneficent law of Nature, that causes her to waste no displays of sublimity or beauty by making them either lasting or common. Before the light of morn is sufficient to make any objects distinctly visible, it displays a beauty of its own, beginning with a faint violet, and melting through a succession of hues into the splendor of meridian day. It remains through the day mere white transparent light, disclosing the infinite forms and colors of the landscape, being itself only the cause that renders everything visible. When at the decline of day it fades, just in the same ratio as substantial objects grow dim and undiscernible, this unsubstantial light once more becomes beautiful, painting itself in soft, tender, and glowing tints upon the clouds and the atmosphere. Similar phenomena attend both the opening and the decline of the year. Morning is the spring, with its pale and delicate tints that gradually change into the universal green that marks the landscape in summer, when the characterless brilliancy of noonday is represented on the face of the land. Autumn is emblemized by the departing tints of sunset; and thus the day and the year equally display the beneficence of Nature in the gradual approach and decline of the beauty and the splendor that distinguish them.

The flowering of the forest is the conclusion of the beautiful phenomena of spring, and summer cannot be said to begin until we witness the full expansion of its foliage. In the early part of the season each tree displays modifications of verdure peculiar, not only to the species, but to the individual and the situation, and hardly two trees in the wood are shaded alike. As the foliage ripens, the different shades of green become more thoroughly blended into one universal hue; and this uniformity, when perfected, distinguishes the true summer phase of vegetation. As summer advances, this monotony increases until near its close. The only trees that variegate the prospect are the evergreens, by their darker and more imperfect verdure, and one or two rare species, like the catalpa and ailantus, which display a lighter and more lively green, resembling the verdure of early summer.

It may be said, however, in behalf of summer, that no other season affords so good an opportunity to note the different effects of sun and shade in the foliage of the woods and fields. The leaves of the trees and grass are never so beautiful in their summer dress as they appear during the hour preceding sunset, when we view them with the sun shining obliquely toward us. All foliage is more or less transparent, and the rays of the sun, made slightly golden by the refraction of the atmosphere, communicate a brilliant yellow tinge to the leaves, as they shine through them. The same effects are not produced by reflection; for if we look away from the sun, the foliage and grass present a much less attractive appearance. A few hours after noonday, before the sunlight is yellowed by refraction, we may study these phenomena more minutely. When we look in the direction of the light, as I have just remarked, we see the least variety of light and shade; for as every leaf is an imperfect mirror, the surface of the forest presents a glitter that throws a glazed and whitish appearance over the green of the foliage. The whole is a mere glare, so that the landscape is almost without expression when viewed in this manner, and all the tiresome uniformity of summer verdure is aggravated. The only relief for the eye comes from the shadows of isolated trees and small forest groups as they are cast upon the ground.

Now let us turn our eyes in an opposite direction. To obtain the best view, we should look obliquely toward the sun. Then do we behold a magnificent blending of light and shade; for every mass of foliage has a dark shadow beneath it, forming a more appreciable contrast on account of the intense brilliancy, without glitter, caused by the illumination of every leaf by the sunlight shining through it. Under these circumstances we can once more distinguish species, to some extent, by their colors. We shall soon discover that trees which have a thin corrugated leaf, without gloss, make the most brilliant spectacle when viewed in this manner. Nothing can surpass the foliage of the elm, the lime, the maple, and the birch in this peculiar splendor. But trees like the poplar, the tulip-tree, the oak, and the willow, having a leaf of a firmer texture and less diaphanous, look comparatively dull under the same circumstances.

I would repeat that the true summer phase of wood-scenery is that which succeeds the flowering of the forest, when all the different greens have faded into one dark shade of verdure. There is no longer that marked and beautiful variety which is displayed before the maturity of the leaves. Summer is not, therefore, the painter's season. It is dull and tame compared even with winter, when regarded as a subject for the brush or the pencil, and especially when compared with spring and autumn. Summer is the time for the observations of the botanist, not for those of the picturesque rambler; for beneath this sylvan mass of monotonous verdure the sods are covered with an endless variety of herbs and flowers, surpassing in beauty those of any other season.

But the flowers are far from being conspicuous. The bright-colored species are not sufficiently profuse to modify a general view of the landscape. Indeed, the uniformly dark shade of green pervading every scene after midsummer is not relieved until near the end of August, when the golden-rods appear and diffuse a yellow lambent hue over the borders of fields and brooksides, multiplying day by day until their colors are universal. The golden-rods are indeed the harbingers of "yellow autumn." Their hues are the dawning of that splendor which from this period gradually overspreads the face of nature.

In a summer forest scene, the evergreen woods are the principal enliveners of its monotony. Even the dingy hues of the juniper and cypress become by position the beautifiers of the landscape, acting as a foil to the deciduous trees, and causing their verdure to be more striking. The homely pitch-pine always pleasantly modifies the drowsy effect of a deciduous wood, as the monotony of sweet music is enlivened by occasional interludes of harsh, if not discordant strains. Such is the effect of scattered groups of evergreens. The sameness of summer forest scenery cannot be as great as I have described it, if there be a goodly share of coniferous trees. Few forest scenes are more striking than a deciduous wood in July, all green and lustrous in the sunshine of noonday, with frequent groups of white pine towering above the general level, and spreading out their summits of dark green foliage like natives of another clime. Imagine, on the other hand, the beauty and effulgence of a little grove of white birches and tremulous poplars with their white and shining branches and trunks, their fluttering leaves, and their airy spray, standing on an elevation that overlooks a gloomy swamp of pine and cypress.

THE OAK.

If the willow be the most poetical of trees, the Oak is certainly the most useful; though, indeed, it is far from being unattended with poetic interest, since the ancient superstitions associated with it have given it an important place in legendary lore. It is not surprising, when we remember the numerous benefits conferred on mankind by the Oak, that this tree has always been regarded with veneration, that the ancients held it sacred to Jupiter, and that divine honors were paid to it by our Celtic ancestors. The Romans, who crowned their heroes with green Oak leaves, entitled the "Civic Crown," and the Druids, who offered sacrifice under this tree, were actuated by the same estimation of its pre-eminent utility to the human race. When we consider the sturdy form of the Oak, the wide spread of its lower branches, that symbolize protection; the value of its fruit for the sustenance of certain animals; and the many purposes to which the bark, the wood, and even the excrescences of this tree may be applied, — we can easily understand why it is called the emblem of hospitality. The ancient Romans planted it to overshadow the temple of Jupiter; and in the adjoining grove of oaks, — the sacred grove of Dodona, — they sought those oracular responses which were prophetic of the result of any important adventure.

To American eyes, the Oak is far less familiar than the elm as a wayside tree; but in England, where many

> " . . . . a cottage chimney smokes
> From betwixt two aged oaks,"

this tree, formerly associated with the principal religious ceremonies of that country, is now hardly less sacred in the eyes of the inhabitants from their experience of its shelter and its shade, and their ideas of its usefulness in all the arts. The history of the British Isles is closely interwoven with incidents connected with it, and the poetry of Great Britain has derived from it many a theme of inspiration.

The Oak surpasses all other trees, not only in actual strength, but also in that outward appearance by which this quality is manifested. This expression is owing to the general horizontal tendency of its principal boughs, the great angularity of the unions of its smaller branches, the want of flexibility in its spray, and its great size compared with its height, all manifesting power to resist the wind and the storm. Hence it is called the monarch of trees, surpassing all in the qualities of nobleness and capacity. It is the embodiment of strength, dignity, and grandeur. The severest hurricane cannot overthrow it, and, by destroying some of its principal branches, leaves it only with more wonderful proof of its resistance. Like a rock in mid-ocean, it becomes in old age a just symbol of fortitude, parting with its limbs one by one, as they are withered by decay or broken by the gale, but still retaining its many-centuried existence, when, like an old patriarch, it has seen all its early companions removed.

A remarkable habit of the Oak is that of putting forth its lower branches at a wide angle from the central shaft, which rapidly diminishes in size, but does not entirely disappear above the lower junction. No other tree displays more irregularities in its ramification. The beauty of its spray depends on a certain crinkling of the small branches; yet the Oak, which, on account of these angularities, is especially adapted to rude situations, is equally attractive in an open cultivated plain. It forms a singularly noble and majestic standard; and though surpassed by the elm in grace, beauty, and variety of form, an Oak of full size and just proportions would attract more admiration.

The foliage of the Oak may be readily distinguished at all seasons. It comes out in spring in neatly plaited folds, displaying a variety of hues, combined with a general cinereous tint. Hence it is very beautiful when only half developed, having a silvery lustre, intershaded with purple, crimson, and lilac. The leaves, when fully expanded in all the typical oaks, are deeply scalloped in a way which is peculiar to this genus of trees; their verdure is of more than ordinary purity; they are of a firm texture, and glossy upon their upper surface, like evergreen leaves. In midsummer few forest trees surpass the Oak in the beauty of their foliage, or in its persistence after the arrival of frost.

Oak woods possess characters almost as strongly marked as those of a pine wood. They emit a fragrance which is agreeable, though not sweet, and unlike that of other trees. They seldom grow as densely as pines, poplars, and other trees that scatter a multitude of small seeds, and, being soft wooded, increase with greater rapidity. The Oak is slow in its perpendicular growth, having an obstinate inclination to spread. It has also a more abundant undergrowth than many other woods, because it sends its roots downward into the soil, instead of monopolizing the surface, like the beech. One thing that is apparent on entering an Oak wood is the absence of that uniformity which we observe in other woods. The irregular and contorted growth of individual trees, twisting in many directions, and the want of precision in their forms, are apparent at once. We do not see in a forest of Oaks whole acres of tall slender trees sending upward a smooth perpendicular shaft, as we observe in a wood of beech and poplar. Every tree has more or less of a gnarled growth, and is seldom entirely clear of branches. If the branch of an Oak in a dense assemblage meets an obstruction, it bends itself around and upward until it obtains light and space, or else ceases to grow without decaying, while that of any soft-wooded tree would perish, leaving the trunk smooth, or but slightly defaced.

TREES IN ASSEMBLAGES.

Open groves, fragments of forest, and inferior groups alone are particularly interesting in landscape. An extensive and unbroken wilderness of wood affords but a dreary prospect and an unattractive journey. Its gloomy uniformity tires and saddens the spectator, after some hours' confinement to it. The primitive state of any densely wooded continent, unmodified by the operations of civilized man, is sadly wanting in those cheerful scenes which are now so common in New England. Nature must be combined with art, or rather with the works of man's labor, and associated with human life, to be deeply interesting. It is not necessary, however, that the artificial objects in a landscape should possess a grand historical character to awaken our sympathies. Humble objects, indeed, are the most consonant with nature's aspects, because they manifest no ludicrous endeavor to rival them. A woodman's hut in a clearing, a farmer's cottage on some half-cultivated slope, a saw-mill, or even a mere sheepfold, awakens a sympathetic interest, and enlivens the scene with pastoral and romantic images.

A great part of the territory of North America is still a wilderness; but the forests have been so extensively invaded that we see the original wood only in fragments, seldom forming unique assemblages. Especially in the Western States, the woods are chiefly sections of the forest, scattered in and around the spacious clearings, without many natural groups of trees to please the eye with their spontaneous beauty. They surround the clearing with palisades of naked pillars, unrelieved by any foliage below their summits. They remind me of city houses which have been cut asunder to widen an avenue, leaving their interior walls exposed to sight. These fragments of forest, and the acres of stumps in the recent clearings, are the grand picturesque deformity of the newly settled parts of the country. But when a wall of these forest palisades, a hundred feet in height, bounds the plain for miles of prospect, it forms a scene of unexceptionable grandeur.

It is chiefly in the old States that we see anything like a picturesque grouping of trees. There the wood assumes the character of both forest and grove, displaying a beautiful intermixture of them, combined with groups of coppice and shrubbery. Thickets generally occupy the low grounds, and coppice the elevations. The New England system of farming has been more favorable to the picturesque grouping of wood, and other objects, than that of any other part of the country. At the South, where agriculture is carried on in large plantations, we see spacious fields of tillage, and forest groups of corresponding size. But the small, independent farming of New England has produced a charming variety of wood, pasture, and tillage, so agreeably intermixed that we are never weary of looking upon it. The varied surface of the land has increased these advantages, producing an endless succession of those limited views which we call picturesque.

When a considerable space is covered with a dense growth of tall trees, the assemblage represents overhead an immense canopy of verdure, supported by innumerable pillars. No man could enter one of these dark solitudes without a deep impression of sublimity, especially during a general stillness of the winds. The voices of solitary birds, and other sounds peculiar to the woods, exalt this impression. Indeed, the grandeur and solemnity of a magnificent wood are hardly surpassed by anything else in nature. A very slight sound, during a calm, in one of these deep woods, has a distinctness almost startling, like the ticking of a clock in a vast hall. These feeble sounds afford us a more vivid sense of the magnitude of the place, and of its deep solemnity, than louder sounds, which are attended with a confused reverberation. The foliage, spread out in a continuous mass over our heads, produces the effect of a ceiling, and represents the roof of a vast temple.

In an open grove we experience different sensations. Here pleasantness and cheerfulness are combined, though a sense of grandeur may be excited by some noble trees. In a grove, the trees in general are well developed, having room enough to expand to their normal proportions. We often see their shadows cast separately upon the ground, which is green beneath them as in an orchard. If we look upon this assemblage from an adjoining eminence, we observe a variety of outlines by which we may identify the different species. A wild wood is sometimes converted into a grove by clearing it of its undergrowth and removing the smaller trees. Such an assemblage displays but few of the charms of a natural grove. A cleared wild wood yields shade and coolness; but the individual trees always retain their gaunt and imperfect shapes.

Artificial plantations display the characters of a grove; but all spontaneous growths are bordered and more or less interspersed with underwood. Hence a limited growth of forest, like a wooded island, surrounded by water or by a meadow, surpasses any artificial plantation as a picturesque and beautiful feature of landscape. The painter finds in these spontaneous collections of wood an endless variety of grouping and outline for the exercise of his art; and the botanist discovers, in their glens and hollows, hundreds of species that would perish in an open grove. Some woods are distinguished by a superfluity, others, like fir and beech woods, by a deficiency of undergrowth, and this differs in botanical characters as well as in quantity, according to the predominant species in the wood. In all woods, however, shrubbery is more abundant on the borders than in the interior. This border-growth contributes more than anything else to harmonize wood and field. It is the outside finish and native embellishment of every spontaneous assemblage of trees.

A wood in a valley between two open hills does not darken the prospect as if it covered the hills, though, if it be continuous, it hides the form of the ground. But when it has come up in scattered groups on a wide plain, without the interference of art, it surpasses every other description of wood-scenery. An assemblage of trees on a hillside is called a "hanging wood," because it seems to overhang the valley beneath it. Thus situated it forms oppositions of a very striking sort, by lifting its summits into the sunshine while it deepens the shadows that rest upon the valley. Wood on steep declivities is an interesting sight, especially if an occasional opening reveals to us the precipitous character of the ground, and shows the difficulties which the trees have overcome in their struggle for life. Some of our pleasure comes from the evident utility of such a wood. We see at once that a rocky steep could not be occupied by any other vegetation, except under the protection of the trees, and that trees alone could resist the force of occasional torrents; that without them the ground would be barren, ugly, and profitless, and difficult and dangerous to those who should attempt to climb it.

THE WHITE OAK AND OTHER SPECIES.

The most important, though not the largest, of the American trees of the Oak family, and the one that is most like the English tree, is the American White Oak. It puts forth its branches at a comparatively small height, not in a horizontal direction, like the white pine, but extending to great length with many a crook, and presenting the same knotted and gnarled appearance for which the English oak is celebrated. Individual trees of this species differ so widely in their ramification that it would be difficult to select any one as the true type. Some are without a central shaft, being subdivided at a small height into numerous large branches, diverging at rather a wide angle from a common point of junction, like the elm. Others send up their trunk nearly straight to the very summit of the tree, giving out lateral branches from all points almost horizontally. There is a third form that seems to have no central shaft, because it is so greatly contorted that it can only be traced among its subordinate branches by the most careful inspection. The stature of the White Oak, when it has grown in an isolated situation, is low, and it has a wider spread than any other American tree.

The leaves of the White Oak are marked by several oblong, rounded lobes, without deep sinuosities. They turn to a pale chalky red in the autumn, remain on the tree all winter, and fall as the new foliage comes out in the spring. The tree may be readily distinguished from other oaks by the light color and scaly surface of the bark, without any deep corrugations. In Massachusetts very few standard White Oaks have escaped the axe of the "timberer," on account of the great demand for the wood of this species. Were it not for the protection afforded by men of wealth to oaks in their own grounds, all the large standards would soon be utterly destroyed. Democracy, though essential to republican liberty, is fatal to all objects which are valuable for their poetic or picturesque qualities. It has no foresight, and no sentimental reverence for antiquity. It perceives the value of an object for present use; but it disdains to look forward to the interest of a coming generation. In regard to nature, what is called progress in America is only another name for devastation. How great soever the political evil of large estates, it is evident that in proportion to their multiplication will be the increased protection afforded to our trees and forests, as well as to the birds and quadrupeds that inhabit them.

THE SWAMP OAK.

The Swamp Oak bears resemblance in many points to the White Oak; but it has less breadth, and abounds in straggling branches growing from the trunk just below the junction of the principal boughs. This gnarled and contorted growth is one of the picturesque appendages of the Swamp Oak, distinguishing it from all the other species, and rendering it an important feature in a wild and rugged landscape. This cluster does not, like the vinery of the elm, clothe the whole extent of the bole, but resembles an inferior whorl of branches below the principal head. Above it, the tree forms rather a cylindrical head, and the principal branches are short compared with those of other oaks.

The leaves of this tree bear some resemblance to those of the chestnut. They are almost entire, and bluntly serrated, rather than scalloped. They are of a slightly reddish green when mature, and turn to a leather-color in the autumn. Trees of this species are at the present time very prominent objects of the landscape in Eastern Massachusetts, where they are very frequent in half-cleared lands that lie only a little above the sea level and contain considerable clay. The Swamp Oak in some favorable soils attains great size; but in New England, though an interesting object in scenery, it is only a tree of second magnitude. The Chestnut Oak is not uncommon around New Bedford and many other parts of New England, but it is not an inhabitant of the woods near Boston.

THE RED OAK.

The Red Oak is the largest of the genus belonging to American woods, and the least useful for any purposes except those of shade and ornament. It is very regular and well proportioned, having a remarkably wide spread, and branches comparatively but little contorted. It is taller than the white oak, and does not branch so near the ground; but it possesses in a high degree that expression of majesty for which the oak is celebrated. The scarcity of trees of this species by our roadsides is remarkable, since they display the union of so many of the qualities which are desirable in a shade-tree. The Red Oak thrives well on a poor soil, and grows with great rapidity; its foliage is very beautiful, and deeply cleft, like that of the scarlet oak, though larger, and its reddish-purple tints in the autumn are hardly inferior. Perhaps the scarcity of oaks in general by the wayside is owing to the peculiar shape of their roots, which extend to a great depth in the soil, and render the trees very difficult to be transplanted. Hence the wayside oaks are such as have come up spontaneously in the places they occupy, and were there when the road was laid out.

THE SCARLET OAK.

The Scarlet Oak in many points resembles the one I have just described. Like the red oak, its branches are regular and comparatively free from contortions, and the quality of its timber is inferior. The leaves are distinguished from those of all other species by their deep sinuosities, being almost like the skeletons of a leaf, the lobes terminating in narrow teeth with long sharp points. This tree is greatly admired in landscape, and on large estates it is constantly preserved as an ornament. Its chief attraction is the bright color of its autumn foliage; but the fine gloss and deep verdure of its leaves in summer are very beautiful. It turns in autumn to a dark crimson, not a scarlet, as the name would imply. It could not justly be named scarlet, save when it is brightened by sunshine, which adds to all crimson foliage a little gold. But as the oaks are very late in assuming their autumnal tints, and are not in their brightest condition until the maples have faded, the Scarlet Oak, when it has attained its full splendor, is the most beautiful tree of the forest.

There are certain trees which we do not highly value in landscape as single individuals, while they attract our attention in assemblages. Our hills, for example, in some parts of the country, are nearly covered with a growth of Scrub Oak, or Bear Oak. They are not ornamental as single trees, and they are prone to usurp the whole ground, excluding that charming variety of shrubs which constitutes the beauty of our half-wooded hills.

THE BLACK OAK.

It is not my intention to enumerate all the species of this genus; but I must give a passing notice to the Black Oak, because it is a common and very large tree in favorable situations. It has been named Black Oak on account of the very dark color of its outer bark; and Yellow Oak, — a name quite as common as the other, — from the yellow color of its inner bark, which produces the *quercitron* used by dyers. It may also have been so called from the yellowish leather-color of its leaves in the autumn, resembling the color of a dry oak-leaf. Many large trees of this species are found in the New England States. In Kentucky it is named Black Jack, and constitutes the principal timber of those extensive tracts called Oak Barrens.

HOMELINESS OF NATURE.

It seems a part of the benevolent plan of Nature to adorn her works, in general, with great frugality, and homeliness is accordingly the prevailing character of her scenery It is, indeed, necessary that whatever is common should be so unattractive as not to deaden our susceptibility to agreeable sensations when we behold scenes of actual beauty. Nature uses it, therefore, only as a luxury; for occasional and sparing adornment, sometimes a little while in profusion, but never making it a lasting feature of any prominent objects. Her ordinary aspects are agreeable, as colorless light to the eye and pure water to the taste, but the one is not sweet and the other not beautiful. She has provided a comfortable state of being for our usual condition; and does not by her ordinary phases keep the mind or the senses in a state of excitement. She takes care that our sensitiveness to the influence of beauty shall be preserved in a healthy state by the general rudeness and sobriety of the landscape.

It is not denied that these homely objects may possess a kind of relative beauty, coming from our idea of their adaptedness to our pleasures and from pleasant associations. Many an ugly scene on the face of the earth may seem beautiful from its power of awakening the pleasures of memory. So keen is this sentiment, that we often with difficulty distinguish homely scenes and objects thus consecrated to our affections from such as are intrinsically beautiful. I rank under the head of intrinsic beauty those qualities only that produce an agreeable and stimulating effect upon the visual organism. In this acceptation of terms the general aspect of nature is homely. And it would be easy to show that such an adjustment of creation is promotive, not only of our general well-being, but that it preserves our capacity to enjoy the sight of all beauty as it comes before us.

Hence as Nature leads up the seasons of the year she presents their unattractive features in the most conspicuous light, and makes their beauties so evanescent that they usually bear the name of harbingers, because their infrequency seems only to warn us of a change. Spring she escorts like a fair maiden garlanded with flowers, binding her brows with lilies and snowdrops, and causing thousands of minute beauties to rise wherever she places her feet. All these soon pass away, seldom remaining long enough to tire us of their presence. Summer bears the horn of plenty, contributing more directly to our physical comfort, but not so deeply affecting the imagination. Summer presents us with occasional outbreaks of splendor, but never wearies the eye by their frequency. Autumn, amid the waning lights of heaven, for a short period wins our admiration by spectacles of unusual splendor. Then for a few weeks the face of nature may be called beautiful. But were this scene of splendor continued through the year, its charming influence would be lost upon us. Hence those æsthetic philosophers who recommend to stimulate the mind and sight with universal ornate scenery would soon render our faculties morbidly dull to beautiful impressions.

But Nature has been economical in her luxurious proffers for sense and appetite. She neither strews the ground with gaudy colors, nor causes wine to flow in streams, like crystal water. She chooses rather to strengthen our perceptions by a cautious frugality, making the common objects of the material world rude, homely, and even repulsive, occasionally revealing some charming scene to captivate the sight, and bearing some agreeable fruit to regale the other senses. Thus, when we stroll among the leafless woods in spring, while all things wear the aspect of desolation, the mind is stimulated by the absence of beauty, so that not a flower appears on a mossy knoll that is not greeted with delight.

Men who live all the year under the continuous splendor of the tropics lose in great measure this healthful susceptibility to the charms of ordinary objects. This may explain why in those countries only where there is an annual revolution of the seasons and a constant change in the phases of nature has civilization made any considerable progress. I am convinced that the homely features of landscape are the proper aliment of the soul: they are its strength and refreshment, and they preserve its healthful tranquillity. The occasional displays of beauty are but as gleams of light from heaven penetrating through our shadows, and revealing glimpses of the great mysterious Source of happiness that lies beyond our ken. There is but little that attracts our attention in the formless clouds that produce the summer rain. But after the rain has passed away, when the sun reappears involved in mist, faintly illuminating the scene, like hope in adversity, the mind is exalted to rapture; and when at length, renewing the ancient covenant between earth and heaven, the rainbow comes forth to signalize the renewal of all blessings, then do we feel the influence of the highest material beauty.

It is idle to say that by the canons of decorative art the landscape cannot be made too beautiful. As well might we say that there cannot be too much fragrance in the air, or too much color on the blue surface of a lake. If the whole ground were covered with brilliant flowers, if trees were all perfect and symmetrical in their shape, and not rugged and homely as now, how sadly would they be wanting in their present attractions! The very pebbles and gravel and broken sods that intercept our progress, and often offend the sight, are as needful parts of the great picture as the most beautiful object that attracts our admiration. Beauty, like a pearl or diamond, derives value from the endeavors we make to find it; and an occasional glimpse produces more pleasure than we should derive from the constant sight of it.

Hence the most brilliant and enchanting colors and forms are chiefly confined to the minute objects of creation, while those most apparent from their magnitude or extent of surface are sober in their hues, or rough and rude in their general appearance. All the exceptions to this law among things of considerable magnitude are such as retain their brilliancy only a few moments. In the forms and hues of the clouds, which are always evanescent, in the frostwork upon our windows, in flowers and fruits, in birds and insects, has nature displayed the most beautiful hues, forms, and combinations. But the rocks that compose the hills and mountains, the general outline of the forest, and the surface of the ground, are destitute of beauty, and are attractive only as they cherish some agreeable sentiment. But Nature does not withhold the charms of color and symmetry from her evanescent forms, however great their magnitude. Hence the incomparable beauty of the rainbow!

In her displays of mere organic beauty, Nature will not bear comparison with art; and if we are more charmed with her scenes, it is because they more powerfully affect the imagination. All would agree, upon drawing a comparison between the beauty of the clouds at sunset and that of the interior of a dome of colored glass, that the latter is the more brilliant and variegated.

But the artificial scene produces only the organic sensation of beauty, while the natural scene exalts the mind with enthusiasm. In the one case, we view a mere artful and splendid arrangement of colors in symmetrical combination; in the other, what seems like an opening of the gates of Paradise. Yet I well remember a time when there was a brilliant and beautiful sunset, the clouds being arranged in the sweetest harmony of colors and forms, showing every gradation of hue from gold and crimson and orange to purple and violet; I had just stepped into the interior of a dome of colored glass, remaining several minutes, with a few friends, studying and admiring the exhibition. When we returned into the open air, the western clouds had at that moment attained their highest splendor. But our sight was so dazed by the intense brilliancy of the colored glass, that the glories of sunset seemed to all the company dull, faded, and without character. Alas! thought I, how by the luxuries of art may we destroy our capacity to be moved by those appearances that would serve to delight the sense and to elevate the soul to heaven!

THE LAUREL.

Of the Laurel, so celebrated in the romance of classical literature, there are only two species in the New England States, — the Benzoin and the Sassafras. But those two shrubs, being deciduous, are not associated in the minds of the people with the true Laurel. They have given this name to the Kalmia, which is evergreen and bears a superficial resemblance to the Laurel of the poets. A curious fact is related by Phillips, in his "Sylva Florifica," of the Laurel, which may not be out of place in these pages. In the Middle Ages, favorite poets, who were generally minstrels, were crowned with wreaths of Laurel branches containing the berries; and this custom was imitated in colleges, when they conferred a degree upon graduating students. "Students," says Phillips, "who have taken their degrees at the Universities, are called *bachelors,* from the French *bachelier,* which is derived from the Latin *baccalaureus,* — a laurel-berry. These students were not allowed to marry, lest the duties of husband and father should take them from their literary pursuits; and in time all single men were called bachelors."

THE SASSAFRAS.

The Sassafras-tree is usually a shrub in this part of the country, abounding in almost all woods, and very generally sought for the pleasant aromatic savor of the bark. Occasionally I have seen the Sassafras growing to the height of a middle-sized tree in Massachusetts, but it rarely attains such dimensions except in the Middle and Southern States. All the large trees in this region have perished, and I have not seen one since my boyhood, when there were many of them. I am therefore led to believe that the changes in our climate consequent upon the general clearing of the forest, whatever their general effects may be, have not been favorable to the Sassafras, which has become extinct as a tree in this latitude.

The Sassafras often attains the height of sixty feet in the Southern States, and nearly forty feet in the country round Philadelphia. The leaves, when young, are downy, very deeply lobed, mucilaginous, and aromatic. The flowers are greenish, inconspicuous, and only slightly fragrant. The berries are of a bright blue color, and are the favorite food of some small birds. On account of its agreeable aromatic properties, the Sassafras became known to the Europeans at an early period, and was very generally employed in medicine. At present it is simply used as an aromatic stimulant. Gerard calls it the ague-tree, and it was believed to be efficacious in the cure of many diseases. There is a tradition that the odors of the Sassafras, wafted from the American shore, led Columbus to believe that land was near, and encouraged him and his mutinous crew to persevere on their voyage.

THE BENZOIN.

The Benzoin is never more than a middle-sized shrub, sometimes, though rarely, attaining the height of eight or ten feet. It is not branching, but sends up its long stems, like some of the dwarf willows, directly from the root, without assuming a tree form. We often find these long branches covered with foliage from the root to the extremity. The leaves are of a handsome ovate form, and are highly aromatic, but differ essentially from the Sassafras in their odor. The berries have been used as spice for culinary purposes.

CLIPPED HEDGE-ROWS.

No art connected with gardening has been so generally ridiculed in modern times as the topiary art, or that of vegetable sculpture. It is certainly not worthy of defence; and yet it seems to me quite as rational to cut out a figure in box or yew, as to shear the branches of a hedge-row to reduce it to architectural proportions. I cannot see why vegetable architecture is any more rational than vegetable sculpture. I cannot see why those persons who admire a clipped hedge-row should object to an "Adam and Eve in yew," or a "Green Dragon in box," nor why those who are willing to torture a row of shrubbery by this Procrustean operation should not be pleased with a "Noah's Ark in holly," or an "old maid-of-honor in wormwood," as described in Pope's satire. Of the two operations, I consider the one that still maintains its ground in popular taste the most senseless. "An old maid-of-honor in wormwood" would at least have the merit of being ridiculous; but a clipped hedge-row is simply execrable, without affording any amusement.

TREES AS ELECTRIC AGENTS.

To a poetical mind there is no exercise more agreeable than that of tracing in the economy of Nature certain trains of causes and effects that seem to represent her as a kind benefactor, aiming to promote the happiness of all creatures. While we treat of the beauty of trees and of their capacity to afford shelter, shade, and salubrity, it is pleasant, while continuing our observations, to find no end to the advantages that flow from them. We have studied them as the beautifiers of landscape, as the sources of vitality and salubrity in the atmosphere, as our shade in summer and our shelter in winter; as the cause of equability, both of temperature and of moisture. We may also discover in them and their branches an infinite number of lightning-rods, presenting millions of points both for the discharge and the absorption of electricity. Trees differ from other plants in this respect only by presenting their points at a greater elevation, where they can act more immediately upon the clouds.

Trees, especially in dense assemblages, may therefore, in frequent instances, be the immediate occasion of showers, by conducting to the earth the electric fluid of the clouds, and inducing that non-electric state which precedes the discharge of rain. This seems to be effected by electric disorganization. An *organized* cloud is an aggregation of vaporous particles, which are suspended in the atmosphere and held in a state of *union without contact.* Being in a similarly electrified condition, they are kept separate by that law of electricity which causes two pith-balls, suspended by threads, when similarly electrified, to repel each other at certain distances. All those clouds that show a definite and organized arrangement, and resemble feathers or lace, are charged with electricity. As they accumulate they lose their symmetrical arrangement, but do not mix, until some object, charged with opposite electricity, comes near them and draws from the mass its electric fluid, when the vaporous particles, losing their mutual repulsion, immediately coalesce and descend in rain.

To illustrate the action of trees in producing showers, we will suppose a dense electric cloud to be passing over a dry plain containing only a few trees. Not meeting with any conducting objects of appreciable force on its journey, it remains suspended in the heavens until it reaches either a large collection of water, or encounters a forest, over which, as over a lake, there rests always, in calm weather, a stratum of invisible moisture, which is a powerful conducting agent. The trees, with their numerous vegetable points, and the vapor that overspreads them, combine their force in drawing down the electric fluid from the cloud passing over, causing the whole mass to descend in showers. The damp stratum of air which, in still weather, rests upon the surface of every large sheet of water, being a powerful conductor, serves to explain a phenomenon often observed in a dry season near the coast. A dense electric cloud is seen to pass over our heads, without shedding a drop of rain, until it reaches the ocean, when the humid air above the waves, acting as a conductor, causes the cloud to part with its electric fluid and to fall in copious showers at the same moment.

Occasionally a similar cloud, after rising in the west about thirty degrees, will be turned from its direct course, and repelled by the dry, heated atmosphere resting on the plain, and, attracted by the invisible cloud of moisture that hovers over the river valley, is seen to take the course of the river in its journey toward the sea. Hence it is notorious that in a very dry time the rivers obtain more showers than the plains, and the wooded mountainous regions more than the open and level country. And we may regard it as a happy accident in the economy of nature, that trees should be the most serviceable in nearly all other respects, hardly less than as electric agents, upon those situations which are of the least value for the purposes of agriculture. Their branches on lofty ridges and elevations, extending near the level of the lower clouds, are like so many lightning-rods on the buildings of an elevated city, and exert a powerful influence in conducting the electric fluid from an overcharged atmospheric stratum, and preventing, in some degree those accumulations that produce thunder-storms. Nature employs this grand vegetable apparatus as one of the means of preserving that equilibrium, both of moisture and electricity, which cannot be greatly disturbed without dangerous commotions.

I have said nothing of trees as a protection from lightning; but there are many curious facts and superstitions on record in relation to this point. "When a thunder-storm threatened," as Suetonius relates, "Tiberius never failed to wear a crown of laurel-leaves, impressed with the belief that lightning never touched the leaves of this tree." The general opinion that certain trees are exempt from the stroke of lightning is very ancient. It probably originated in some religious ideas of their sanctity, and men in more enlightened times have endeavored to explain it by philosophy, instead of rejecting it as fable. It was affirmed by Hugh Maxwell, an American writer, that lightning often strikes the elm, the chestnut, the oak, the pine, and less frequently the ash; but it always evades the beech, the birch, and the maple. Captain

Dibdin remarks, in a letter to Alexander Wilson, that in the forests of Virginia the pines, though taller than the oaks, were less frequently injured by lightning, and considers them pretty secure when growing among oaks. These accounts by different writers are too various and contradictory to be of much value in aiding us to discover the truth. It is probable that the partial exemption of certain trees from the stroke of lightning, if any such accounts be true, depends on their size and shape. A tall tree in an assemblage would be more exposed than the others. It may also be supposed that if a tree has a regular ramification, smooth and straight branches and trunk, it is better formed for a conductor, and that it would be more liable to receive a charge of the fluid. But all these opinions are probably of the same character with those respecting the antipathy of serpents for certain trees, — traditionary notions which are hardly worthy of investigation. The opinion of the ancients concerning the immunity of the laurel was probably derived from their idea of its sanctity as the tree which was dedicated to Apollo. At the present day there exists in Italy a similar notion concerning the white grapevine. Some of the peasantry of that country are accustomed to twining its branches around the head and waist as a protection from a thunder-stroke.

Trees are generally believed to protect a house adjoining them from lightning; on the contrary, it is known that men and animals seeking refuge under a tree in an open plain are in greater danger than outside of it. The lightning is therefore probably conducted by the water passing down on the surface of the branches and trunk; for if the tree itself were the conductor, the lightning would pass through the trunk into the ground, and, like a lightning-rod, act as a protection to objects near, but not in contact with it. Dr. Franklin thought the safest place a few yards distant from a tree, and a little outside of its widest spread. It is unsafe to stand under the drip of a tree, which might convey to the person an electric charge. It was the opinion of M. Arago, that trees overtopping houses at small distances cannot be regarded as affording sure protection, like a properly adjusted lightning-rod; but he admitted that when a storm passes over a forest it is decidedly enfeebled. The forest certainly diminishes the power of a thunderbolt. The security derived from trees attaches principally to large assemblages. Though a house may receive but little protection from a few tall trees standing near it, it is not to be denied that a village or hamlet is rendered more secure by adjoining woods.

THE GROUND LAUREL.

THERE is only one Epigea in this country,—a very fragrant and beautiful species, creeping close to the ground, and bearing dense clusters of pearly flowers, edged with crimson. The flowers are not unlike those of some of the heaths, though of larger size. It grows abundantly in many parts of New England, particularly around Plymouth, and in various localities from Canada to Georgia. It is a creeping shrub, occupying dry knolls in swampy land, and growing along on the edges of the swamp upon the upland soil. The leaves are almost round, evergreen, light-colored and slightly russet, partially overlapping the dense clusters of flowers, that possess a great deal of beauty and emit an odor like that of hyacinths.

No plant has more celebrity among our people than the Ground Laurel, the earliest of all our wild flowers. I cannot consent to apply to it the common unmeaning name of "Mayflower," thus associating it with the fetid Mayweed, and falsifying its character by an anachronism that assigns to the month of May a flower belonging to April. The name of Mayflower, as applied to the Epigea, means nothing except what is false. Almost all our early flowers belong especially to the month of May. This is distinguished from them by appearing almost alone in April. Its popular appellation is a plain misnomer; and as an apology for it, the name is said to have been given to it by the Pilgrims, in commemoration of the ship that brought them to this country. I cannot believe the Pilgrims ever took any notice of it. Mayflower is a name that originated with some ignorant people, who could not think of any better name than the one it bears in common with fifty other species.

THE BEARBERRY.

THE Bearberry is a more common plant, and more elegant in its foliage, with less conspicuous flowers, than the ground laurel. This plant covers extensive tracts on the borders of woods and partially under their protection. The foliage, resembling that of the box, has always been admired, and nothing makes a neater or more beautiful covering of the turfs which it adorns. The Bearberry is a native of both continents. It abounds in light sandy soils, forming a frequent undergrowth of a pitch-pine wood. The berries are eaten by quails and robins in winter, when they can seldom find any animal food except a few dormant insects.

THE CHECKERBERRY.

THE Checkerberry is peculiarly an American plant, well known by its pleasant aromatic flavor, its shining evergreen leaves, its delicate white flowers, and its scarlet berries. There are no wild fruits so attractive to young persons, from the time they begin to redden in the autumn, and all through the winter, when the ground is open, until they are seen hanging on the vine with the blossoms of spring. Indeed, this fruit is not perfected until it has remained on the bush during the winter. The severest cold has no effect upon it; and the berries increase in size, after the spring opens, until they become as large as strawberries.

This plant is very abundant in all woods in New England, and seems to be confined to no particular soil or situation. Indeed, I doubt whether another woody plant can be found so generally distributed throughout the New England forest. If it has any preferences, they seem to be the lower slopes of wooded hills and mountains. But I have seen it in all locations where it can enjoy the protection of trees, in evergreen as well as deciduous woods; for though the leaves of the pine prevent the growth of any considerable underwood, the Checkerberry is always abundant in the openings of a pine forest.

LILY-PONDS.

SOME of the most delightful prospects are comprised within a narrow compass; and such, indeed, are the average of those scenes which have been selected for the painter's canvas. When we ascend a high mountain, we generally observe that the most enchanting views are beheld from some point not far from its base, where the objects of attention are circumscribed by surrounding eminences. A valley of small extent, inshrined among wooded hills, if it be not so exhilarating as a scene of wider view, is certainly more satisfactory and more picturesque. Here the imagination finds scope for agreeable exercise, without the weariness produced by a view of illimitable space, and the consequent seeking after something beyond our ken. Nature does not surfeit her intelligent creatures with scenes of beauty or grandeur. She economizes her wealth and her resources, and makes no attempt, like ambitious men, when improving her works, to dazzle the sight with uninterrupted splendor. She has opened many little valleys among the hills, to collect within them a greater amount of beauty than she assigns to ordinary places; and to crown them with the highest attractions she has placed a lily-pond in their centre, to present at one view all that is charming in landscape, either to the painter or the poet.

All the beauty of nature and all the life of the woods gather spontaneously about a lily-pond. Here assemble the water-birds of various plume, attracted by the fishes and the plants that are gathered around the shore. The

singing birds of the wilderness find here both sun and shelter, and a vegetation more fully stocked with insect life. Nowhere is there so much animation, apart from human abodes, as on the grassy banks and wooded eminences that surround it; nowhere is there so much beauty outside of the domain of human art. The variegated wood-duck finds seclusion here among the tall rushes, and subsistence near the fertile shores, abounding with water-cresses and other aquatic herbs. And the youthful angler, standing on the shore, watches with delight the little spotted plover, as it runs nimbly upon the lily-pads, then casts his line over beds of aquatic flowers as sweet as a garden of hyacinths.

If we thread the footpaths that make their labyrinthine course around the pond, we shall be struck with the profusion of beauty with which nature has encompassed it. These paths, the chance work of cattle, — picturesque artists, unconscious of their art, — constantly entice us into some dew-bespangled nook, fringed with mosses and garlanded with ferns, or lead us up some gentle eminence that affords a view of the pond and its irregular margin, and, through the openings of the wood, a peep into the neighboring landscape. Everywhere do we meet with pleasant surprises, where the precipitous banks, intersected by rush-bearing inlets and covered with wood, conceal all intimation of the approaching view.

To one who is fond of quiet amusements, there is no greater luxury than to float along the shores of the pond in a little skiff, and survey the scenes it opens to our view without wearisome toil. From a boat we see only the perfect sides of the trees, where, meeting with no impediment, they spread out their full and natural proportions. Here every outline is perfectly shaded, with a pencilling peculiar to nature, varied with many fantastic forms, without uniformity, and yet without abruptness. Nature uses her different vegetable species to produce her wonted pictures, — the elm and the birch for her flowing and drooping lines; the oak, with its gnarled and sturdy branches, for grandeur; the maple for beauty, at certain seasons; and the hemlock, with its silvery-spangled foliage, for brilliancy and grace. All these, and many others, she has planted around the pond, and filled up the space between the ground and the lower branches with an undergrowth of sweet-scented shrubs, so that from the bosom of the lake the boatman might imagine himself in a scene of enchantment.

Nature seems to have the same affection for a lily-pond as for old waysides in the country, which have not been trampled by the too frequent passing of travellers; and on their borders, with equal hand, she groups her vegetation in the same anomalous dispositions as we observe in the forms of clouds. Sometimes the pond sends out a branch, forming a shallow, where beauty gives place to wildness. In these dank meadows Nature creates many grotesque forms of vegetation. Giant rushes raise their spears, half buried in water; and the tupelo-tree by its twisted and fantastic growth adds a tinge of romance to the view. Here variety and uniformity, wildness, rudeness, and beauty are blended in a charming manner unattainable by art. I speak of those ponds that remain undisturbed by human hands, having neither been made the location of ice-houses, nor modified to suit the taste or the pride of the owner of some neighboring villa. I speak of them as they come from the hands of Nature, or as modified only by the simple operations of the rustic farm.

Many of these beautiful ponds have been appropriated by traders, or spoiled by country-seats that substitute the beauty of art for that of spontaneity and destroy all their original features. But there are thousands of them still quietly sleeping in the forest, unshorn of their original attractions. On the boundaries of these virgin waters, Nature still presides, where Art has not introduced her affectations, nor Pride desecrated a single scene by her baleful ornaments. There is not, during all the season, a day when the plaintive song of the veery may not be heard from the shore proclaiming itself the chief chorister of the woods, from the time of the flowering of the rhodora until the clethra scents the groves in midsummer; while the fairest flowers, the clearest fountains, birds that dwell in sacred retreats never profaned by the plough, trees that for centuries have spread their harps to the tuneful gales, roses that have annually offered the purest incense to the skies, ambrosial herbs that deck the ground with their verdure, then perish and offer their leaves as a balm to the sick,— cupbearers of incense to the dewy morn and even, — all rise, and bud and bloom, and scatter their fragrance, and weave a warp of beauty in a friendly ambuscade around the dwelling-place of the water-lilies.

The angler, if he be either a philosopher or a naturalist, can deeply feel the charm of all these objects. I can imagine no man more happy than one who, after passing the greater part of the day in the occupation that affords him a livelihood, resorts to these secluded retreats to obtain that tranquillity which cannot be found in the bustle of commerce, to breathe the incense rising to heaven wherever the flowers are bathed in dew, and to gaze upon the charming array of beautiful things that sparkle around the altar of Nature. Bright gem of the forest, fixed under the brows of these wooded hills for the baptism of the votaries of Nature into her inner sanctuary of delights! Above thy glassy wave the happy angler may watch the shifting forms of the clouds as they pass languidly over its mirrored surface, while zephyrs laden with the perfume of violets hover round him and fan him with their wings. Among these scenes, how beautiful are the shadows that rest upon the silvery pond, and how musical the sounds that come up mysteriously from the woods and dingles!

Our lily-ponds, for the most part, are surrounded by hills, that form a basin for their waters, and become the principal source of their replenishment. Not in the deep waters, however, nor under the steep banks, but in the shallows, near the outlet, do the water-lilies congregate, fixing their roots in the alluvium, and extending their long stems upward to the length required for raising the bud to the water's brink. As soon as it has gained this height it is ready to become a flower, which expands about the third hour after sunrise, and remains open until the shadows of the woods are cast upon it in the afternoon. If at any hour the sky should be overcast with clouds, the flowers close their petals, yielding their honors to the more homely yellow lily, the pontederia, and the nodding sarracenia upon the shore. All the seasons have garnered around these waters a portion of their stores; and both to the naturalist, who studies the character and habits of animate and inanimate objects, and to him who chiefly observes nature's beautiful aspects, the lily-pond is a page written over and over with myriads of lines, letters, and pictures, without confusion, and perfectly legible to those who, spurning the pleasures of a luxurious life, resort here to live nearer to nature and to happiness.

THE BEECH.

THE Beech is a common tree in all our woods, where it is distinguished by the length and size of its smooth clean shaft, which is often perceptibly ribbed or fluted. In dense assemblages these columns, rising to the height of sixty or seventy feet, are very striking, and the more so when the land is covered entirely with Beech timber. The suckering habit of this tree and its vigorous constitution are the important cause of its predominance in any tract that is occupied by it, and the close matting of leaves that covers the ground under a beechen wood prevents any abundance of undergrowth. The same inconvenient habit is the cause of its rareness in dressed grounds. George Barnard says of the English Beech: "In no tree are the decaying hues of autumn more beautiful than in the Golden Beech, its foliage changing from green to the brightest orange, then to glowing red, and eventually to a russet brown, in which state the leaves remain on the tree through the winter." The leaf of the American Beech, on the contrary, is remarkably dull in its autumnal tints. It turns to a rusty yellow in the autumn, gradually fades to a leather-color, and drops from the tree near midwinter.

The style and spray of the Beech, as observed in its denuded state, are worthy of particular study. The lower branches of the tree are generally very long and rather slender. They take an almost horizontal direction when they start from the tree, but soon make a curvature by turning regularly upwards, and causing a peculiar primness in their general appearance. Every small twig also turns upwards, pointed with elongated leaf-buds, resembling so many little spears. The terminal branches, forming the spray, are very numerous and slender, and remarkably beautiful. The Beech, when in full leaf, is seen to the best advantage where it skirts the edge of a wood, if it has grown up there since the original clearing. In that situation we perceive the elegant sweep of its branches, and the upright character of its leaves, each leaf pointing obliquely upwards in the direction of the spray, instead of hanging loosely in all ways, like the foliage of the large-leaved poplars. Deciduous trees have generally a drooping foliage, and the want of this habit in the Beech gives it a very lively appearance. The heaviness attributed by Gilpin to the English tree is not observed in the American Beech; on the contrary, it is remarkable for a certain airiness, seldom putting forth its branches in masses, but in such a manner that every spray may be traced by the long upright rows of leaves.

I should hesitate in saying that on cultivated ground, and as a standard, the Beech would display those qualities which are most admired. It is chiefly interesting by the woodside, or skirting the banks of a stream. The stiffness of its foliage renders it ungraceful as a solitary standard. It may be remarked, in its favor, that it differs so widely in its ramification from other deciduous trees as to add a pleasing variety to any miscellaneous assemblage of species. I can easily believe that it is not a favorite resort for birds; for its branches are too long and slender for their convenience, and its foliage too thin to give them a feeling of seclusion. If I were to plant a grove of beeches, I would select the crumbling banks of water-courses, where the trees would bind the fragile soil with their roots and cover the banks and the hillside with a beautiful wood and an agreeable shade.

The tendency of the Beech to produce mosses and lichens upon its trunk and branches has been observed by the earliest writers. It is also a matter of common observation among woodmen. No such growth, however, is seen upon beeches that stand alone or in an open grove. These parasites are generated by the dampness of a thick forest; and they attach themselves equally to the bark of other trees in the same damp situations, but cannot adhere to it if it be rough or scaly. The smooth bark of the Beech, and of the red maple while it is young, permits such plants to foster themselves upon it, and adhere to it without disturbance.

THE RUSTIC LANE AND WOODSIDE.

NATURE is greatly indebted to Art for many of her attractions, if it has not been exercised for the purpose which is effected by it. We see this not only in wood-paths, which all will agree are the most delightful parts of a wood, but in many other operations of a rude agriculture, more especially in the rustic lane. It is no matter whether the lane be bordered by trees and shrubbery, or only by a plain wooden fence or loose stone-wall, provided for several seasons it has been entirely neglected. It must have been long enough under nature's spontaneous action to restore that condition of the turf that precedes cultivation, to green the borders with ferns and mosses, and to gem their velvety heaps with anemones and violets. The nice trimming and weeding which are generally apparent in all the paths and avenues of a country-seat or a model farm deprive them of the attractions of the rustic lane. No matter how many flowers are cultivated in the borders of one of these trim avenues, it is, after all, only an exhibition of splendor and luxury. It delights the eye, but it cannot win the heart. It is only a conservatory of elegance; it is not a paradise.

If we follow the course of any rustic lane which has not been improved, bounded by a rude fence of any kind which will form a support for the plants that come up beneath it, we see the climbing and creeping plants in their unrestrained freedom and beauty. If in the course of our walk we meet with a rude shed or any building old enough to be overgrown with mosses and incrusted with lichens, its walls are sure to be covered either with the climbing sumach or the Virginia creeper; for these plants seem designed by nature as the native embroidery of all neglected places and buildings. On many accounts, the most interesting plants are the climbers and creepers. Whether it be that we associate them with the idea of dependence on their part and of protection on the part of the tree or other object that supports them, or whether their ascent may suggest the idea of motion and progression, causing them to resemble a living creature, they never fail to interest the spectator, and to fill his mind with many poetic images.

The Virginia creeper possesses all the advantages of the English ivy, save that it is not an evergreen. But its deciduous character is not to be regarded as a defect, since if it were an evergreen it would want its annual attractions of scarlet and crimson that distinguish it in autumn. In this particular it is not surpassed by any production of the American forest, except the red maple. These colors render it very conspicuous in October, when it surrounds the trunks and branches of some of the tallest trees with its garlands of crimson, hiding them under its own splendid frondage. There is not a rustic lane where it is not seen creeping over the fences and mixing its glowing tints with other wayside plants. It is particularly luxuriant by the woodside; for though it is common in the deep forest it grows feebly and is deficient in leaves until it gains the summits of the trees. It needs the broad eye of day, and prospers only upon trees that stand outside of a wood. No other climbing plant is so generally used in New England as a drapery for houses and fences, taking the place occupied in Europe by the ivy. Many old houses are covered by it, and many an old stone-wall is completely enveloped in its foliage.

The poison ivy, or climbing sumach, is the only rival of the Virginia creeper in our woods. It is even more common in open fields, and though less luxuriant, surpasses it in the beauty of its leaf. It is a very pertinacious parasite, adhering very closely to the object that supports it, with its innumerable rootlets, but sustaining life only by communication with the soil. The growth of this plant is discouraged on account of the liability of many persons to be injuriously affected by its poisonous properties. Those who are not familiar with wild plants are generally unable to distinguish the poison ivy from the Virginia creeper. Their general appearance and habits are nearly the same, but their leaves furnish a sure mark of distinction. They are compound in each; but those of the Virginia creeper are in fives, those of the poison ivy in threes, without exception.

As we pass along the rustic lane, where it is involved in deep shadow by a dense growth of shrubbery and vines we see the woody nightshade adorning the mass with its singular halberd-shaped leaves, its dark blue flowers with a golden centre, and its pendent clusters of scarlet fruit. I know but few plants of which so little has been said that possess a greater share of beauty. There is a common prejudice against the woody nightshade, from its supposed poisonous qualities, and from our habit of identifying it with the deadly nightshade of Europe. If our plant has some poisonous qualities, they are not of a dangerous character. All parts of it may be bruised and handled with impunity, and its berries are so nauseous to the taste and smell that they are not liable to be eaten.

In the wild hedgerows that skirt our fields and farms, made up of viburnum, elder, cornel, hazel, and wild rose-bushes, the woody nightshade, in company with the glycine, contributes greatly to the interest attached to these flowering thickets. What excites my surprise is that so few persons praise this modest little climber. How would its varied foliage, interwoven with that of more luxuriant plants, the deep but contrasted colors of its flowers and fruit, and its constant presence in the borders of all wet fallows, attract the admiration of a painter who, imbued with a love of nature equal to his love of art, should attempt to paint a New England stone-wall with its many native accompaniments!

A more conspicuous climber, and more common by the woodside than by the rustic lane, is the bitter-sweet. It is seen climbing over trees, not attaching itself by rootlets or tendrils, but twining round its supporter, like the morning-glory. It is often fifteen or twenty feet in height, covering some unfortunate tree with its own dense foliage, and finally causing it to perish by excluding light and air from it. This plant is well known to simplers, who have named it bitter-sweet, from the mingled sweet and bitter of the scarlet and orange-colored berries which they collect for medical use. I cannot learn that they contain any medicinal virtue; but it is well understood, in these days, that the possession of decided efficiency renders any medical substance unpopular. All popular remedies are physic only to the faith; hence the incomparable virtues of saffron and elder-flowers, whiteweed and everlasting!

We are prone, when thinking of plants merely as ornaments of nature, to forget that the fruit-bearing shrubs and vines have in general anything to recommend them except their fruit. It will be admitted that very many of these plants are deficient in beauty; yet I will confess that I have often admired the different species of bramble, which are so common in the rustic lane and woodside, trailing over fences and abrupt elevations, or hanging down from projecting cliffs, and exposing their clusters of red, black, and purple fruit. Our common species are not remarkable for elegance or beauty, but the country waysides would look bald and cheerless without the simple decoration afforded by these plants.

Among the trailing species of bramble, one of the most important as a natural ornament of lanes and field-borders, is the dewberry, or evergreen blackberry. It is very abundant on the edges of woods, where the trees are thin and scattered, and in pastures covered with low shrubs, where it may be recognized by its small, elegant, and shining leaves. These in protected situations remain green all winter, becoming slightly impurpled as spring advances. The dewberry covers with its close network of trailing branches the virgin turf which has been left undisturbed in the borders of lanes and wood-paths. When the soil has been repeatedly turned by the plough, this little inhabitant of the primitive sods gives place to a larger species, that trails in a similar manner upon the ground, and bears an excellent fruit.

The only native species of bramble which is admired for the beauty of its flowers, but not so common in fields and lanes as in old gardens, is the flowering raspberry. It is so called from the size of its large crimson flowers with a yellow disk, resembling a dark red single rose. The leaves of this species are not pinnate, like the leaves of other species of bramble, but palmate, resembling the leaf of the striped maple. We sometimes find it in a shady nook, concealing itself under a stone-wall, and seldom in company with other shrubs. The delicacy of its habit unfits it to contend with its more hardy congeners, and it is soon driven away from its retreat by the ingress of other species.

I have not yet spoken of the grapevine, which, if not very ornamental in gardens, where its beauty is marred by excessive pruning, cannot be surpassed in a certain kind of suggestive or relative beauty. Hence the pleasure it affords us when we see it on the borders of woods, hanging its purple clusters of fruit over some placid stream from the summit of an alder, or hiding the rudeness of a neglected building with its broad foliage. There is hardly an old road or rustic byway in the interior of the country which is not festooned by wild grapevines, and some of the most delightful arbors on old country roadsides are formed by these vines, trellised upon an ancient apple-tree or drooping birch.

When a green by-road passes over a wet meadow and crosses a brook under a natural arch formed by overhanging alders fastened together by creeping vines, the shade afforded by this arbor is greatly heightened by a twining canopy of clematis, or virgin's bower, climbing over the trees and shrubs, always keeping on the outer surface, and supporting itself by tendrils. We often pass through copses of shrubbery completely overspread by this vine, rendered conspicuous when in fruit by multitudes of little silken and feathery tufts, which are far more beautiful than its flowers. There is not much beauty in this plant, and I attribute the interest attached to it chiefly to its poetical name and the romantic history of the European virgin's bower.

THE CHESTNUT.

Many admirers of trees place the Chestnut before the oak because it is a taller tree with a proportional spread and denser foliage. A remarkable peculiarity in the style of its foliage is its radiated tufts, giving it a similar appearance to that which is so apparent in the horse-chestnut. But we observe an important difference between the two, — while the radiated tufts of the horse-chestnut are distinctly separated by spaces, those of the Chestnut seem to be involved in a general and more indistinct mass of foliage. A notion prevails in some parts of Europe, that this tree should not be planted near dwelling-houses, "because the flowers emit a powerful and disagreeable odor, which is offensive to most people." I have not observed any such odor from the American Chestnut.

In general form and proportions there seems to be no specific difference between the English and the American chestnuts. On this continent it is a majestic tree, remarkable for the breadth and depth of its shade; but it is seldom cultivated by roadsides. It displays many of the superficial characters of the red oak, so that in winter we cannot readily distinguish them. The foliage bears some resemblance to that of the beech, but displays more variety. The leaves are long, lengthened to a tapering point, and of a bright and nearly pure green. Though arranged alternately, like those of the beech, on the recent branches, they are clustered in stars, containing from five to seven leaves, on the fruitful branches, that grow out from

the perfected wood. When the tree is viewed from a moderate distance, the whole mass seems to consist of tufts, each containing several long pointed leaves, drooping divergently from a common centre. From this centre the aments of the male flowers come out in a similar way; and their bright silvery green, glistening upon a mass of darker foliage, always attracts attention at the time of flowering.

The Chestnut is ranked among the largest of our forest trees, sometimes in favorable situations attaining a height of nearly eighty feet. When growing isolated on a plain, its diameter is sometimes equal to its height. The Chestnut has a rather loose ramification, being in this respect inferior to the red oak, which it resembles. Its larger branches are numerous, but the spray is coarse, the terminal branches being fewer and more straggling than those of the oak. This tree is therefore not comparable in beauty with the oak when divested of its leaves. The Chestnut is a classical tree, being mentioned very frequently in the works of the Greek and Roman poets, who were familiar with it.

THE SENTIMENT OF ANTIQUITY.

In America, the few remains of antiquity which have been discovered belong to so remote an age, and are so insignificant compared with those of Europe and Asia, that they do not characterize the land nor affect the newness of its appearance. Even nature, outside of its geological structure, seems less ancient here than on the old continent, where we find a greater number of trees which are survivors of a remote period of the past. Trees do not attain their greatest age in a forest, where their crowded condition is unfavorable to longevity; and if any in this country have become very aged, they do not display the sturdy and venerable appearance that marks old standard trees. Hence, in the productions both of nature and of art, America is new in comparison with Europe and Asia. Though its mountains and valleys, its rocks, meadows, and river-beds, are as old as those of the Eastern continent, it is not hoary with the ruins of ancient grandeur, nor shaded by trees that for centuries have spread their umbrage over the same field and roadside.

Though the European would look in vain in his own country for those features that charm the admirer of our half-cultivated landscapes, he is not so often offended by disagreeable contrasts presented here in opposition to our wood-scenery in the works of a flashy architecture. In an old country, save those parts which have been changed by recent improvements, its artificial works have been sobered and mellowed by time. Structures originally showy have lost their glitter; they wear a look of repose and sobriety, and, like rocks, seem to be a part of the ground they stand upon. A magnificent building, which was at first too highly ornamented with the gilding of vanity to be justly valued for its intrinsic merits, in the course of time becomes sobered into an expression of simple grandeur. Many such edifices exist in Europe, and yield to its artificial landscapes a venerable appearance which is entirely wanting in those of America.

Individuals of a poetic and thoughtful turn of mind are generally more attached to the old than to the new, and the improvements they are willing to make are such as are not destructive of the historic remnants of a past century. People of this character among our inhabitants are lovers of Nature, who presents to their sight many of the semblances of antiquity. A wood which we have always frequented may be the only object in our village that wears an ancient look, except the rocks and hills. I am aware that very little of this sentiment pervades the active classes of American society, who are so eager to increase their wealth by new enterprises, that every change is delightful to them if it precedes a commercial adventure. I have seen men in raptures over the demolition of some of the most charming scenes of their boyhood, on beholding them laid out into house-lots, and advertised for sale. They are so deeply interested in advancing the price of "real estate," that they do not think of the regret with which, at some future day, they may witness the desolation that has followed. These sacrifices are constantly becoming necessary to the wants of an increasing population; but if our people were more deeply imbued with this sentiment of antiquity, many interesting objects would be preserved which are needlessly destroyed.

The women of America have generally more culture than the men, except among the literary classes, and feel more regard for the preservation of any object that derives value from imagination or sentiment. If there be a dozen persons in a village who would save from threatened destruction an old building or a venerable grove, these few are chiefly of the fair sex. The regard we feel for such things is generally proportioned to our culture; not to our intellectual power, which is quite another thing. But men of active and practical habits are prone to despise a sentiment that lies too deep for their sensibility, and refuse to preserve any relic of the past that will not improve the commercial value of their own property. The wanton sacrifice of trees is often condemned in connection with the building of new roads. But trees are not the only valuable objects in a natural landscape. The spade and the pickaxe may do more injury than the axe of the woodman to the face of the country, and of a kind that is irreparable. Collections of shrubbery upon certain picturesque eminences, fern-clad rocks projecting from the brow of a hill and overhanging the roadside, have the charm of venerable ruins joined with the freshness of living vegetation. By grading all these to one dead level or slope, the scene is despoiled of its beauty and deprived of its picturesque associations.

It has often been asserted that the scenery to which men have been accustomed from their youth produces an effect on their character corresponding with its state of rudeness or cultivation, its tame and smooth or abrupt and mountainous surface. The influences of society, however, must greatly counterbalance these effects, which are, after all, very problematical. It is hard to believe that the wild scenes of nature would turn men into savages, except as they deprive them of education and of intercourse with civilized people. On the other hand, it will be admitted that they encourage, by their peaceful solitudes, any meditative habit of mind which certain individuals among the inhabitants might possess. Our people are educated to admire only the new by their constant familiarity with new buildings, new villages, and new cities; and the tendency of all these objects is to foster an excessive love of glare and ornament. If they have any regard for antiquity, it is in most cases a passion for certain relics of ancient art which have been imported for their fashionable value. In such cases the taste for antiquities proves to be nothing more than a rage for novelties.

A writer in the "London Critic" has remarked that the Americans ought to be devout worshippers of nature, inasmuch as "twenty steps will take the meditative man into the wilderness." But this remark will apply only to "meditative men." The active members of our population have come by habit to regard the natural condition of the country as the great obstacle in the way of their material prosperity, and they feel no affection for objects that must be destroyed to promote their thrift. We cannot love anything which is a hindrance to our success in any darling project. The most devout lovers of nature in this country are among those who were brought up in the oldest towns and villages, where the primitive forest has been succeeded by one of sparser growth, who prize their trees and groves as property increasing in value, and look upon them with pleasure as a part of their valuable possessions. A familiarity with the new prompts to enterprise, speculation, and mechanical invention. Old scenes and objects encourage thought, stimulate the imagination, and foster the poetic sentiment. The lovers of the new are pleased with art chiefly as it contributes to show and splendor; the lovers of the old, as it serves to render the scenes of this earth better subjects of inspiration. Hence the genial influence of an old country full of venerable ruins. On this continent art is almost entirely new, and has been used rather for the display of art than to awaken any noble sentiment.

The pleasure afforded a lover of antiquity by the scenery of a new country must be awakened chiefly by the objects of nature; for art is disagreeably vapid and ostentatious where the wealthy inhabitants are chiefly ambitious to surpass each other in the parade of their resources. If the wild and rude character of the landscape were destroyed, if the spontaneous woods were despoiled, and nothing remained but a general baldness, nature would afford but little relief from the glare and insipidity of ornamental art. Yet I cannot feel that the venerable buildings of an old country full of antiquities would make amends for the absence of the wild and spontaneous scenes of nature. Not many districts on the old continent can be so attractive as New England, which more than any other land displays that charming intermixture of the wildness of nature and the beauties of civilized art which is apparent in all the interior. And these features it will always retain, so long as the man who tills the soil is the owner of it, and every laboring farmer is an independent yeoman.

THE HICKORY.

The Hickory, including several species, is very generally distributed over this continent, but is found in no other part of the world. It is distinguished from the walnut by its foliage and general habit of growth, by the smaller number of leaflets on the leaf-stem, and by their darker color and firmer texture. The aments of the Hickory are in threes, and the outer shell of the fruit opens at four angles when it is ripe; the aments of the walnut are single, and the outer shell of the nut is undivided. The two trees differ also in their general appearance. The Hickory rises to a greater proportional height, with less length and spread of the branches, the lower ones being higher from the root of the tree and smaller than those of the walnut. Many of the trees are flattened at the top, and take a cylindrical form, when they approach to any regularity; but their outlines are more frequently irregular, displaying frequent gaps, and presenting several distinct masses of foliage.

The Hickory, therefore, when full grown, has seldom much elegance, and little of the beauty of grace and symmetry. Its picturesque qualities are its sturdy habit, its great height, its dense and dark green foliage, its approach to a cylindrical shape, and its general eccentricity of growth. I have never seen a Hickory with long spreading branches like those of the butternut, nor with neat and prim foliage like that of the ash. The different species are so common in all the southern parts of New England as to form a notable arboreal feature of our landscape. In Massachusetts we see them following the lines of the old stone-walls, having come up from nuts planted by squirrels on the strip of land around the borders of the fields. We are indebted to this fortunate circumstance for thousands of beautiful and valuable trees, which, but for this narrow border of neglected land, would not have been allowed to "cumber the ground." The trees that originated in these borders had ample room to expand, assume their normal shape, and acquire their full dimensions; and as we see them running upwards with but little width, we may consider this to be their natural style of growth.

Hickories are abundant on fertile slopes, near brook-sides, and on rocky hills that abound in clay and yellow loam. They do not prosper on light, sandy soils, and are not found in bogs. They are even a better indication of a fertile soil than the oak. The shellbark alone drops its leaves before they are tinted in the autumn. The most remarkable species in New England are the shell-bark, the fignut, the white hickory, and the bitternut. These four have nearly the same outward characters. They are, indeed, so much alike that the shellbark alone is readily distinguished by the exfoliation of the outer rind of its bark as soon as it has come to fruit-bearing. The bark of the other three species is channelled or furrowed, like that of the ash. The fruit of the fignut is fig-shaped; and as the epithet *ficiformis* was very early applied to this species, it is evident that the vulgar name of pignut is a corruption of the true name, which ought to be restored.

Had the old painters been acquainted with the Hickory, they would have admired it beyond most other trees. The peculiarities of its shape are remarkable. The breaks in its foliage cause that variety and irregularity of outline which are generally regarded as picturesque qualities. I see, while I am writing, directly before my window, a tall Hickory, standing on an elevation that makes the sky its only background. It is tall and narrow in its shape, and its head is divided into five distinct masses of foliage, separated by a considerable opening. Two of these masses are on the right, and three on the left, the highest making a flattened top, projecting over the right side, and hanging down in a large flowing mass. Yet this tree is perfectly normal in its proportions, for I can discover no marks of mutilation in any part of it.

The spray of the Hickory, like other trees with pinnate leaves that bear a large seed, is coarse; but its alternate branching gives it variety, and takes away some of that heaviness so disagreeable in the spray of the ash. All its branches are liable to be twisted, because they cannot be broken by the wind, and these contortions often extend throughout the ramification of the tree. It puts them forth from a central shaft, that usually extends to the summit of the tree, and, being small, they are often bent down very considerably by the weight of their fruit. The geographical bounds of the Hickory are the southern parts of New Hampshire on the north, Tennessee and North Carolina on the south, and the shores of Lake Erie on the west. The wood of the Hickory is exceedingly hard, heavy, and tough, and is in America the symbol of courage and firmness.

RELATIONS OF TREES TO TEMPERATURE.

NOT long since, in one of my rambles in Essex County, Massachusetts, which is one of the most open and cultivated sections of the country, I entered a little valley near the sea, comprising about fifty acres of well-cultivated land, surrounded by a sort of amphitheatre of hills, which were covered with a dense forest of pines and firs. It was occupied by an intelligent farmer, whose careful observation had taken note of many things which are overlooked by the generality of his class. He remarked that his seed-time and harvest were several days earlier than on the farms in the open country, and that he had crocuses and tulips in his garden, on the south side of the surrounding wood, so early as to astonish his neighbors in the outer world. In regard to the relative temperature of the woods and of the open plain in summer, he remarked that it varied according to the time of day or night. The woods were cooler than the open country, in clear, calm weather, from about nine o'clock in the morning until near noonday; after this time the heat in them increased more rapidly than in the open country, and at the time of dew-fall it was greater in the woods, and continued so during the early part of the night. If the sky were cloudy, not much difference could be perceived at any hour in the temperature of the two situations. In cold and windy weather the woods afforded a comfortable shelter, and this shelter made them apparently warmer, even when the thermometer would indicate no difference.

The theory of my rustic friend contains the general results of all that science has yet discovered in relation to the temperature of woods. But the effects of clearing the forest are so different in different situations as to have given origin to a multitude of theories. This diversity of opinion, however, comes from a partial observation of facts, without their qualifying circumstances. On a hot summer's day we sprinkle our floors with water, for the purpose of cooling the air of the room. But how can it produce this effect, when by evaporation it carries heat from the floor into the very air that is cooled by it? The fact is easily explained. The greater coolness felt when the air of the room is saturated with the moisture evaporated from the sprinkled floor might not be exactly indicated by the thermometer. The sensation of coolness is caused by the increased power of the air to conduct the heat rapidly from our persons,—the effect of its greater humidity. By the same law we may explain why, after a few clear cold days in the winter, if a south-wind arises, we feel as if the cold were greater, because this wind, while it raises the temperature, charges the air with invisible moisture.

The coldness of the atmosphere over grassy meadows when the sky is clear, after the decline of the sun in summer, is a matter of common observation. As this phenomenon is most evident on the clearest nights, it has given rise to the notion that the moon cools the night air. In our rambles after sunset, we have all felt these constant changes of temperature, which are remarkable when walking over an uneven road, the degree of heat corresponding nearly with our altitude. When we occupy high ground, the air is warm and dry; as soon as we descend into a valley, we feel a sudden chill. These differences are not observed on a cloudy night, or when a clear brisk wind is blowing. But in a calm state of the atmosphere, as the lowest stratum of air contains the greatest amount of moisture, its capacity for retaining heat is proportionally diminished. Consequently the heat from the ground is radiated with great rapidity through this damp stratum of air, while the higher strata remain unchanged in their temperature. Indeed, it has been found by experiment that while the greatest heat at noonday in calm summer weather is very near the surface of the ground, yet after dew-fall the highest temperature is several feet above this surface, increasing in altitude for some hours after sunset.

The action of a wood checks this radiation in the early part of the night. Like clouds in the evening, the trees form a canopy of foliage over the ground, and thereby retain the heat many hours after it has escaped by radiation in the open plain. According to these laws of the radiation of heat, a longer time would be required to cool a tract of forest than an equal area of open space, down to a given point. But, on the other hand, a proportionally longer time is required to raise the temperature in the woods to a given point. Hence it is still a question among meteorologists whether the mean annual temperature of a large tract of country is higher or lower when covered with forest than when generally open and cleared. The sun acts with greater force upon an open country; but the radiation of heat is greater in the same ratio during the sun's absence.

In considering the effects of clearing, travellers have often overlooked the important advantages of protection afforded by woods to agricultural crops. Even if the mean annual temperature of a country be the same after it is cleared as when it was covered, it may at the same time be too cold for certain plants which were formerly its common productions, because there are no woods to protect them from the winds by day or from the cold caused by excessive radiation at night. Palestine, two thousand years ago, was a well-wooded country, and all the fruits of the sub-tropical climates were raised there to perfection by its ancient inhabitants. The date-palm, the fig-tree, and the olive grew there and bore fruit abundantly. Palestine is now a treeless country, and the same fruits are incapable of enduring its climate; yet recent observations have demonstrated that its climate is not colder than it was in the days of the kings of Israel. But as the country has been despoiled of its forests, these sub-tropical fruits are deprived of their natural conservatories, and cannot be raised without great labor and expense in preparing artificial protection for them. Let the forests be restored to the hills and mountains of Palestine, and, though the temperature of its summers were not increased, the fields would be protected by these forests from the winds, and the tender fruits, thriving under their protection, would again become abundant.

The principles involved in these and similar facts form a distinct branch of meteorological science, and would require a volume for their illustration. I have only hinted at some of the general conclusions. It is evident, indeed, that the same objects that serve to protect us from cold may in an equal degree protect us from heat. The woodcutters will continue their labor in a deep forest without discomfort on a winter's day, when they could not endure the intense cold of the open country. The earliest flowers of spring, however, are found neither in a wood nor in an open meadow, but under the protection of a wood on its southern border, in little openings that are exposed to the beams of the sun.

THE BUTTERNUT.

THE walnut includes two species in this country, the Butternut and the black walnut, both trees of considerable note and importance. The Butternut is a well-known tree in the Northern States, cultivated to a great extent in rural villages, but not very abundant in the forest, from which it has probably been extirpated for the beauty and value of its wood in cabinet-work. It is everywhere seen in the enclosures of farm-houses, where it is valued for its fruit and admired as a shade-tree. It is not so tall as the hickory, and differs from it in general shape, as I have already remarked, subdividing itself into several large and equal branches, and seldom extending a central shaft above the lowest point of subdivision. It is a tree of wider spread but thinner foliage than that of the hickory. Its pinnate leaves are long, with a great number of leaflets, and of a light and rather mellow green. It resembles the black walnut in its botanical characters; but the fruit of the Butternut is more elongated, that of the black walnut being nearly globular.

Every one is familiar with the Butternut-tree. Its fruit being more easily obtained than that of the hickory, and ripe at an earlier period, the tree is generally plundered before the time for gathering it. The outer rind is pulpy, and full of a bitter sap that blackens the hands when pressed out by cracking the nuts in a green state; for the kernel is ripe while the shell is still green. This stain may be removed by any fresh vegetable acid; and for this purpose boys generally procure the leaves of sheep-sorrel, with which they rub the stains from their hands, and after washing in soft water it is found to be entirely removed, if no soap has been used. I am not sure that painters would see much to admire in this tree; but to a native of New England it is so pleasantly associated with juvenile feasts of nuts in the early autumn, gratuitously strewed by the green wayside, and with the simplicity of country life, that it is difficult to see in the form of this tree anything we do not admire. If its foliage is thin, its proportions are handsome and symmetrical, and when in its prime there is no tree that better adorns a rustic enclosure. The Butternut puts forth its leaves about a week earlier than the hickory. It is common in all the New England States, especially on the Green Mountain range, from the northern parts of New Hampshire to the Sound.

THE BLACK WALNUT.

The Black Walnut is common in all the United States below the latitude of Long Island. It is especially abundant in Pennsylvania, and is also found singly and in small scattered groups in New England. It is a larger and more hardy and rapid-growing tree than the English walnut, but it bears an inferior fruit. This tree does not differ from the butternut in general characters, but it is of greater height and more majestic in appearance. It has very long pinnate leaves, of a pure untarnished green and a warmer look than the darker foliage of the hickory. Both trees produce an elegant wood for cabinet-work, but that of the Black Walnut is preferred, though the wood of the butternut is nearer the color of mahogany.

THE WHORTLEBERRY PASTURE.

THOREAU relates that he once thought of whortleberrying as an occupation for a livelihood. This was said in a quaint and paradoxical humor, but there are multitudes who can sympathize with the feelings that prompted his remark. As a quiet outdoor amusement, it is not surpassed either by angling or botanizing; and I cannot see why the whortleberry field should not have its Izaak Walton as well as the lily-pond or the trout-stream. The freedom enjoyed in the open pasture, the simple and honest people whom we meet there, the tiresome, but still agreeable and emulative task of picking the fruit, are only a fraction of our enjoyments. The chirping of various insects, and their constant sportiveness among the bushes; the motions of birds and the plaintive melody of the wood-sparrow, which is tuneful nearly the whole month of August, — prepare us to be cheerful and delighted with all things. The cattle feeding carelessly upon the hillsides, the scattered groups of trees and the cool shadows they cast upon the green turf, the sweetness of the air, our unrestrained rambling, the precipitous rocks that intercept our way only to disclose a bower of raspberries protected by their walls, the mossy seats under umbrageous pines, the countless wild flowers on every knoll, the pleasant sensation of rest after weariness and of coolness after the heat of exercise and weather, all combine to render the whortleberry pasture a field of delight surpassing all that is written of gardens of orange and myrtle.

The whortleberry is peculiarly an American fruit;

though a few species are common in Middle and Northern Europe, they are in no part of the world so abundant as in North America. The whortleberry tribe of plants form a conspicuous feature of New England landscape, especially near the coast. No single species has been domesticated, though any one of them would well reward the labor of the cultivator if the fruit could not be obtained from the fields. Their fruit is well known to the inhabitants of the Eastern States. Very little has been written upon it, and few persons are aware of its importance to the inhabitants of North America. Botanists make no generic distinction between the whortleberry and the blueberry; but we may distinguish the two at once by their different flavor, and not by their color. The whortleberry is less acidulous, less mucilaginous, and contains a harder seed than the blueberry. The flowers of the two species differ as widely as their fruits: those of the blueberry are large and white; those of the whortleberry are greenish, tipped with red, smaller and more contracted in the mouth. There is no family of plants that runs into a greater number of varieties in a wild state; but I have never seen one that seemed to possess the characters of the blueberry and whortleberry combined. With regard to their colors it may be remarked, that while there are blueberries which are black, there is no whortleberry which is purely blue.

It may truly be asserted that if the cherry and the whortleberry, with all their varieties, were to become extinct, the want of the latter would be most painfully felt by the mass of our population. We were not taught by the Europeans to appreciate the value of our wild fruits. "In Scotland," said one of a company of Scotch girls whom I met in a whortleberry field, "we have no wild fruits. All our fruits are in gardens." In this country, where whortleberries are so common as to be found in all wild lands that are not densely wooded, their fruit constitutes one of our staple productions, of greater value to us than even the cranberry, except as an article of export. During about three months, from the first of July to the last of September, millions of bushels of whortleberries are consumed in this part of the country. People are often deceived by measuring the importance of any article according to its commercial value. Hence the whortleberry pastures are called "waste lands." But were these lands deprived of their products of wild fruit, the want of it would be a grievous affliction to the community. How many poor families earn their livelihood in summer by gathering whortleberries for the market! How many delightful excursions does this fruit-gathering annually afford to the children and youths of our land! The robin, the waxwing, and other birds that consume our cherries, would be diverted from the orchard and the garden by a good supply of fruit from the bushes of an adjoining field; and our cultivators might prevent their depredations by planting the different species by the sides of their fences and in all open situations which are not adapted to tillage.

As an object in the landscape and a field for the botanist and student of nature the whortleberry pasture is worthy of study and full of attractions. This scenery, with all the spontaneous mapping of its beds of shrubbery, its groups of trees, its tussocks of mosses and ferns, its little green hollows spangled with flowers, and its projecting rocks covered with brambles, all intersected widely by the smooth greensward, is peculiar to New England. In the Southern States the whortleberry-bushes are more promiscuously scattered, and are not seen in this delightful grouping, forming with the trees, fruits, and flowers a true symbol of the beneficence of nature. A genuine whortleberry pasture is one of the most beautiful of gardens, — a modern Vale of Tempe, a true Eden, — inasmuch as it is without culture; and abounds from early spring till waning autumn in the most interesting shrubs and flowers of our clime; in August and September sparkling with clusters of shining black and azure berries, and possessing a value which only a New-Englander knows how to prize.

The whortleberry pasture consists chiefly of upland, extending out occasionally into a level meadow, but generally of a hilly and uneven surface, covered with groves and coppice. The pasture must have been fed many years by cattle to acquire its distinguishing features. Without the grazing of these animals the ground would be evenly covered with vines and bushes. The cattle, while feeding upon the grass, consume many of the young plants which have not become woody, and in their irregular course gradually produce this grouping in a manner which is entirely inimitable by art. Hence in an old field the scattered beds of shrubbery, with greensward between them, might be compared to a map of islands, the grass being represented on the map by the water and the bushes by the land; the greensward sometimes widening into a broad expanse of verdure, and then beautifully intersected by intricate masses of shrubbery.

In the lands surrounding the older townships only do we see the whortleberry pasture in the perfection of this picturesque grouping, laid out according to the geometry of nature. In the new settlements the bushes are mixed with trees and stumps in the clearings, and have not acquired any arrangement. But if a whortleberry field has long been pastured by cattle that seldom browse upon the shrubs, the different kinds of vegetation stand in beautiful groups of a thousand various forms, like the figures on tapestry. The rocks that lift up their gray heads, sometimes with smooth flat surfaces, sometimes in lofty protuberances, covered with liverworts and patches of variegated lichens and mosses, and fringed on their edges with diminutive shrubs, form no unimportant part of this peculiar scenery. In every old pasture the different kinds of shrubs are more or less distinctly arranged into groups; some, for example, consisting chiefly of bayberry, others of roses or perhaps of brambles. But in general the plats consist of a promiscuous variety of species, in which some one predominates. One of the most common of these social plants is the sweet-fern, universally prized for its fragrance, at the very name of which we are inspired with pleasant recollections of youthful wanderings. The lambkill is especially prone to form exclusive assemblages, and the most beautiful individuals, when in flower, are generally on the outside of the group.

But there is no end of the smaller plants that spring up everywhere, some in the open space, others under the protection of a tuft of sedge-grass or a broad-leaved fern. The sweet-scented pyrola is abundant in all shady thickets, and the cymbidium and arethusa decorate the low grounds among the nodding panicles of quaking-grass and the spreading flowers of meadow-rue. The loosestrife, with its long pyramidal spikes of yellow flowers, is always conspicuously grouped in the low grounds, side by side with similar plats of low swamp-roses or crimson-spiked willow-herb. But the most attractive flower in the whortleberry pasture is the red summer lily, — the cynosure of the happy children who assemble there, the queen of the meadow, and the delight of every rambler in the coppice.

The man who thinks of nature only as a field for the display of magnificent art may sneer at these rustic scenes and their native ornaments. But pride cannot make unadorned nature contemptible, nor can the grandeur of a princely estate deprive its occupants, if their culture equals their wealth, of the interest with which they behold a field covered with spontaneous vegetation, or a simple rustic farm. From the opening of spring until the fall of the leaf, the whortleberry pasture is a garden full of the fairest flowers and the most healthful fruits. And if Great Britain's isle had been covered with whortleberries, like our New England hills, these fruits would have been celebrated in English poetry, like the fruit of the vine and the olive in the poetry of Greece and Rome.

WHORTLEBERRIES AND HUCKLEBERRIES.

We may vulgarize a word by associating it with the market. The wild pastures abound in summer with well-known fruits, some of jet and some of azure. We go out with a few friends and gather them with flowers, for present amusement. These fruits are *Whortleberries*. This is their poetical and their botanical name, the one that is associated with all the beautiful things that cluster in the same field. These fruits are also gathered for the market, and exposed for sale with cucumbers, new potatoes, and squashes. They are now *Huckleberries*. Shelley has defined poetry to be the art "that lifts the veil from the hidden beauty of the world, and makes familiar objects be as if they were not familiar." This is done partly by a choice selection of words; and whenever a common thing is known by two names equally euphonious, we should always select that which is not in commercial use. We should say Whortleberries if we are writing an essay or a poem about them, and Huckleberries if we are going to buy a few of them in the market. The usages of the market in other matters ought to be excluded from literature. In commerce, for example, fishes are fish; in natural history fish are fishes.

THE HAZEL.

> "Now let us sit beneath the grateful shade
> Which Hazels interlaced with elms have made."
>
> *Virgil*, Eclogue V.

THE Hazel, under which Menalcas invites his brother-shepherd to sit, is a tree of considerable size, while the American hazels are mere shrubs, seldom overtopping a rustic stone-wall. The Hazel among the Romans, like the olive among the Jews, was regarded as the emblem of peace; and this estimation of it was transmitted to the people of a later period. Hence, in popular works of fancy on the language of flowers, this is recorded as its symbolic meaning; and in ancient times a Hazel rod was supposed to have power of reconciling friends who had been separated by disagreement. These superstitions connected with the Hazel, and more particularly the one relating to the Hazel rod, named the Caduceus, assigned by the gods to Mercury as a means of restoring harmony to the human race, probably gave origin to the divining-rod, which was first made of Hazel and afterwards of the witch-elm. It is remarkable that in America this use was made of the hamamelis, a very different plant in its botanical characters, and hence called the Witch-Hazel.

There are two New England species, both delighting in the shelter of rude fences, and producing their flowers before their leaves. They are distinguished chiefly by the shape of their fruit. The common Hazel is the one most generally known. In this the shells or husks that enclose the nuts are of the same round shape, growing in a cluster, and each invested with a calyx like that of an ordinary flower. The Beaked Hazel is a smaller bush and frequents more solitary places than the other. "The calyx enclosing the nut, densely hispid and round at base, is contracted like a bottle into a long narrow neck, which is cut and toothed at the extremity." The whole nut with its envelope resembles a bird's head and beak. A dry sandy loam is the soil generally occupied by the Hazel. Along the old roads that pass over dry sandy plains, that border many of the river-banks in the Northern States, the Hazel, growing in frequent clumps, forms in some of these locations the most common kind of shrubbery. When we see a pitch-pine wood on one side of a road, the cultivated land on the opposite side is usually bordered with a growth of Hazels.

Both species are particularly worthy of protection and preservation. They produce a valuable nut without our care; they are ornamental to our fields and by-roads; they feed the squirrels and shelter the birds, and they add a lively interest to natural objects by their spontaneous products. The Hazel is associated with many pleasant adventures in our early days, with nut-gatherings and squirrel-hunts, and with many pleasant incidents in classical poetry. The Hazel has been a favorite theme of poets, especially those of the Middle Ages. In the songs of that period are constant allusions to the Hazel-bush, probably from its frequency in natural hedgerows, and its valuable fruit.

THE BUTTON-BUSH.

NOT much has been written of the Button-bush. We hear but little of those shrubs that do not readily admit of culture, and are not susceptible of modification by the arts of florists. The Button-bush is confined to wet, solitary places; indeed, it may be considered a true aquatic, as it grows in most cases directly out of the water. It is associated with the complaining song of the blackbird, whose nest is often placed in the forks of its branches, and it accompanies the ruder aspects of nature. It is far from being an elegant plant; and the little beauty it possesses belongs to the perfectly globular shape of its heads of flowers, which are nearly white. It is generally seen bordering the sluggish streams that flow through the level swamps, and often forms little islets of shrubbery in the middle of a sheet of water.

THE CLETHRA.

AFTER the flowers of the azalea have faded, we are attracted in like situations by a similar fragrance from the Clethra, or Spiked Alder, remarkable as one of the latest bloomers of the American flowering shrubs. It bears its white flowers in a long spike, or raceme, somewhat like those of the black-cherry tree. The Clethra, when in blossom, is not destitute of elegance, and it is valuable for the lateness of its flowering. The foliage of this plant is homely, and its autumnal tints are yellow, while the prevailing tints of our wild shrubbery are different shades of red and purple. It is found in wet and boggy places, where it is very common, displaying its floral clusters as late as the fourth week in August. This shrub, when cut up for brushwood, is called the "Pepper-bush" by the fishermen of our coast, from the resemblance of its roundish fruit to peppercorns. The picturesque attractions of the Clethra are not to be despised, when its long racemes of white flowers are seen projecting from crowded masses of verdure on the edges of the wooded swamps.

A SUMMER NIGHT IN THE WOODS.

When the decline of day is plainly perceptible in the lengthened shadows of the trees and the more refreshing coolness of the atmosphere, many rare birds, that since morning have been silent, begin their songs anew. Evening comes not unattended by the same captivating splendors that lead up the Morn, and the same melodies that herald her approach. As she descends from her pavilion of crimson and amber to spread her twilight over the earth, calling down the gentle dews from heaven, and bringing refreshment to the drooping herbs, the heavens show forth their gladness in the myriad hues of sunset, and all animated nature raises a shout of music and thankfulness. But there is a pensiveness in the melodies of evening that sweetly harmonizes with the sober, meditative hour; and the same birds that in the morning pour out their melodious lays as from hearts full of rejoicing, now whisper them in accents more subdued, like the quiet breathing of the winds as they are wafted over the sleeping flowers.

Just before the sun declines, the thrushes, which are true forest warblers, are very tuneful, and continue to sing until dusk. The note of the little veery is the last to be heard, and when his song has ceased the night may be said to have commenced; though, even after this time, the sweet notes of the vesper-bird are occasionally poured out from some station in the open field. But in our woods, at this season, silence does not immediately ensue. A restlessness prevails among the feathered tribes, as if they were yet unprepared to renounce the pleasures of the day. At intervals, for the space of an hour after dusk, an occasional note of complaint is heard in the thicket from different birds, — a shrill chirp from some of the little sylvias, the mewing of the catbird among the alders, and the querulous smack of the red thrush.

Sometimes for several minutes hardly a voice from any creature is heard, and the rustling of the night-wind through the tremulous leaves of the poplar, or its moaning among the high branches of the pine, resembling the murmurs of distant waters, are the only sounds that meet the ear. But this dreary stillness is not of long duration. The droning flight of the beetle, and the whirring of various kind of moths that are busy among the foliage of the trees, are the accompaniments of a summer night, suggesting to the fancy the passing of a ghost, and filling the mind with many mysterious conjectures. Sometimes the owl, on his soft silken wings, glides along with stealthy and noiseless flight, and we are soon startled by his peculiar hooting, — a sound which I can imagine must be terrific to the smaller inhabitants of the wood.

At midnight, in general, the stillness of the winds is greater than by day, and the gurgling of streams is heard more distinctly amid the general hush of nature. Sounds are now the most prominent objects of attention; and every noise from distant places booms distinctly over the plains and hollows. We are affected with something like a superstitious feeling at night, that disposes us to listen with solemn attention to every sound that we cannot immediately apprehend. While absorbed in our revery, the night-jar, as he flies invisibly over our head, occasionally twangs his wings on a sudden descent through the air, in pursuit of his aerial prey, making a sound that to the superstitious, who are unacquainted with the bird, is fearful and perplexing. The first time I heard this sound, which resembles the snapping of a viol-string, was in my school-days, when walking with three of my comrades at midnight on a solitary turnpike road. Not knowing the cause of it, we were affected with a peculiar sensation of awe, which was not relieved until daylight revealed to us the birds still circling over our heads.

Often, while thus affected with a sensation of mystery, and in an interval of stillness that is almost sublime, all serious emotions will be put to flight by a sudden chorus of bull-frogs from a neighboring pool. These sounds, in themselves inharmonious, are so intimately allied with the sweetness and quiet of a summer night in the woods that they seldom fail to excite pleasure. In the course of our midnight saunterings, when we are near any collection of water, the shriek of the common green frog is heard frequently; and the trilling voice of the toad, so continual by day, occasionally breaks the silence of night. The common tree-frog, the prophet of summer showers, seldom heard by day except in damp weather, keeps up a constant garrulity during all still nights in the month of June.

There is no perfect stillness on a summer night. There are gentle flutterings of winds that nestle in the foliage, mysterious whisperings of zephyrs, and humming of nocturnal insects that hover around us like spirits, and seem to interrogate us about the cause of our presence at this unseasonable hour. We catch the floatings of distant sounds, mellowed into harmony by the intervening space, and hardly to be distinguished from the noise made by a dropping leaf as it comes rustling down through the small branches. The stirring of a little bird, as he preens his feathers upon a branch of a tree, uttering an occasional chirp; a little quadruped leaping suddenly through the underwood and secreting itself hastily among the herbage, — are trifles that add cheerfulness to the solemn quietude of night.

I am supposing the night to be perfectly calm; but how calm soever it may be, now and then a breeze will pass fitfully overhead, and the trees will shake their fluttering leaves in the wind. An unbroken stillness may immediately follow, save at intervals a whisper is heard from some unseen object, as if something that has life were watching your motions, or you had obtained a faint perception of sounds from the invisible world.

Among the affecting circumstances attending a night in the woods, I must not omit to mention the sound of bells that proclaim the flight of time. Their sounds add solemnity to the hour, while they afford a pleasant assurance of the nearness of human dwellings. But the single stroke that tells the hour of midnight, repeated at short intervals from different villages, is peculiarly solemn and impressive. We then feel that we are under the very meridian of night, and that darkness is our only protection. The effect of this single toll upon the mind at such a time cannot be described.

I have spoken only of sounds, but they are at midnight hardly more impressive than sights which affect us the more on account of their indistinctness. The swarms of fireflies whirling and darting about in the lowlands are almost the only creatures that can be seen, save now and then some night-bird, as it passes like a dark spot over the half-luminous sky. But these little sparks of insect life do not aggravate the impressions made by darkness. There is nothing about them that excites the imagination or exalts the feelings. One can easily imagine the terror with which the glaring eyes of the jaguar must be beheld by the midnight sojourner in the South American forest. The eyes of the owl, as seen through the trees, might produce similar impressions; but in our quiet woods imagination is the source of all the terrors that might arise from common objects.

The night would afford no mean employment to the naturalist, if he could observe the midnight operations of the still wakeful part of animated nature. There are many nocturnal insects which, though not easily discovered in the darkness, are then in motion, hovering among the foliage, or seeking the open blossom-cup of some flower of the night. At this time only can the active habits of these creatures be observed, when even the deep shadows do not protect them from the bat, the owl, and the goat-sucker, who nightly destroy thousands of these beautiful insects, leaving their torn wings and elegant plumage in the green forest paths, or lodged upon a leafy branch, marking the place of their destruction.

As real objects are but faintly seen, by the same cause the phantoms of darkness are made visible. There are many things in the obscurity that assume dubious and formless appearances, and excite the curiosity, blended with some apprehension. The branches are pictured like the forms of birds and quadrupeds on the sky, and every passing breeze seems to wake them into life and motion. A beam of light appears on the plain, or a shadow on the hill, reminding you of the dusky form of a ghost as it glides half visibly among the indistinct forms of the trees. On a dark night almost all objects are ambiguous. The trees that stand near the borders of streams cast faint shadows upon them, often mistaken for some real objects resting upon their starry surface. Everything that moves reminds you of a spirit; and many are the unintelligible forms that stand around, nodding their heads, and, as it were, beckoning to some kindred monster. You feel as if they were aware of your presence, and were consulting together how they should regard your intrusion into their dusky haunts.

As night draws near its close, we begin to long for the morning; and the crowing of cocks from some distant farm-yard affords a pleasant relief to our weariness and an assurance of the nearness of dawn. The little hairbird, that utters his trilling note at intervals throughout the night, is heard more frequently. At length an occasional twitter from the birds all around us announces that morning is visible. Nature always gives signs of an approaching change, and morning dawn and evening twilight have their respective harbingers; and she usually accompanies them with peculiar sounds from the elements and from animated things. Thus by the croaking of the tree-toad at noonday she augurs an approaching shower, by the chirping of the green nocturnal treehopper she proclaims the approach of autumn; but the birds are Nature's favorite sentinels, whom she employs to herald the morn.

If we now take our stand on an eminence where we can obtain a clear view of the eastern horizon, a luminous appearance may be observed, forming a semicircle of dim whitish light around the brows of Morning. If a thin veil of clouds overspread the arch, the tints will be dark in proportion to their distance from the hidden source of light. Imagine it divided into circles. The inner one will be of a light yellow, the next a tint of gold; beyond that is orange, and as it extends outward it passes through a gradation of vermilion, crimson, purple, and violet, until it melts into the azure of the firmament.

THE WESTERN PLANE.

WHEN journeying through the older towns of New England, the melancholy forms of the ill-fated Planes attract our attention by their superior size, and still more by the marks of decay which are stamped upon all. This appearance is most remarkable in the early part of summer; for the trees are not dead, but some hidden malady caused the first crop of foliage to perish for several successive years. The trees, after putting forth a new crop of leaves from a second growth of buds, had not time to ripen their wood before the frosts of winter came and destroyed their recent branches. This disaster was repeated annually for ten or fifteen years, causing an accumulation of twigs at the extremities of the branches, making a broom-like appendage, and greatly deforming the spray of the tree.

The Western Plane, or Buttonwood, is a well-known tree by the waysides in New England and in the forests of the Middle and Western States. It belongs to a genus of which there are only three known species, and this genus constitutes a whole natural family. It may, therefore, be something more than a fanciful hypothesis, that all its noble kindred have perished and disappeared from the face of the earth, with other plants of a distant geological era, and that the three remaining species are destined to share the same fate, as signalized by the mysterious fatality which has attended both the Western and Oriental Plane. The Buttonwood is remarkable for its great height and magnitude, its large palmate leaves, and its globular fruit. The foliage is rather sparse, of a light, rusty green, and resembles in many points that of the common grapevine. Near the insertion of every leaf, and a little above it, is a stipule forming a plaited ruff that encircles the growing branch. These ruff-like appendages are among its generic marks of distinction.

"The Buttonwood," says Michaux, "astonishes the eye by the size of its trunk and the amplitude of its head. But the white elm has a more majestic appearance, which is owing to its great elevation, to the disposition of its principal limbs, and the extreme elegance of its summit." He considers the Buttonwood "the largest and loftiest tree of the United States." He mentions one growing on a small island in the Ohio River, which at five feet from the ground measured forty feet and four inches in circumference; and he found another on the right bank of the Ohio that measured, at four feet from the ground, forty-seven feet in circumference, or nearly sixteen feet in diameter, and showed no marks of decay. He states that the Buttonwood is confined "to moist, wet grounds, where the soil is loose, deep, and fertile, and it is never found upon dry lands of irregular surface."

It was probably the rapid growth and great size of the Buttonwood that caused our ancestors to plant it so extensively as a shade-tree. It rises also to a great height before it sends out any branches, thereby affording the inmates of houses the advantage of its shade, without intercepting their prospect, and without interfering with passing objects when planted by roadsides. But these noble trees, so conspicuous and so thrifty thirty years ago, have been slowly perishing from some mysterious cause which no theory can satisfactorily explain. It is generally supposed to be connected with a want of hardihood in the constitution of the tree, that renders it unable to endure all the vicissitudes of a Northern climate.

In England the same misfortune has fallen upon both the American and Oriental Plane. The late spring frosts are mentioned as the probable cause of the phenomenon, though there is but little resemblance between our climate and that of England. This tendency of the two species has prevented the general planting of them for shade and ornament. English writers give their preference to the American Plane, which they assert equals the other in size, and surpasses it in beauty of foliage. In England the American Plane has frequently attained a very great magnitude. Selby mentions one which, at forty years from the time it was planted, measured a hundred feet in height. The specific differences between the two Plane-trees consist chiefly in the size and shape of their leaves, those of the Oriental Plane being smaller, and more deeply lobed or divided into segments. Both species have the same habit of annually shedding their bark, leaving the trunk with a smooth and whitish surface.

BEAUTY IN NATURE.

BEAUTY is any quality in an object which, through the medium of the sight, affects the mind, the passions, or the senses with immediate pleasure. If we carefully examine the objects that possess this quality, we shall be able to separate them into two classes, — one affecting the visual nerves with an organic sense of pleasure, resembling sweetness to the palate or fragrance to the smell; the other having a similar power of exciting agreeable sensations of another kind, or some delightful sentiment or affection. The first is true organic beauty; the other, relative or moral beauty. But the latter is so complex that it is difficult in many cases to determine what the property is that produces it. Color seems to be the principal cause of the organic sensation of beauty; beautiful forms derive more of their power from the expression of certain ideas associated with them. Beauty, indeed, is any visible quality in an object that causes a passionate desire to look upon it for the sake of the pleasure it excites. But there is a great deal both in nature and art which is called beautiful that deserves this epithet only as a figure of speech.

The beauty of landscape in general is purely relative. Nature is for the most part homely in her features; and to those who possess a dull imagination, many of her scenes, which a man of feeling would describe with rapture, are positively ugly. The love of the ornate distinguishes those who possess a great deal of culture with very little imagination, and who seek in what they might call the beauty of landscape some visible quality that gratifies their sense of wealth and pride. There is a certain charm in the ornate finish of dressed grounds that affects all persons with pleasure when it is viewed in the town or in its suburbs; but it clashes with the simplicity of a genuine country scene.

Much of the beauty we perceive in nature delights the eye by promising gratification to some other sense, especially if it is blended with pleasing forms and colors. The sight of a golden apple, or peach with cheeks of crimson and purple, would delight the eye, even if we did not associate these fruits with their gratefulness to the palate. But their fitness to afford this pleasant gratification heightens their beauty; so that it would be difficult to determine, when we look upon a peach-tree laden with its highly colored fruit, whether the sight or some other sense is the most agreeably affected. When passing through shrubbery, our eyes are attracted by the scarlet berries of the woody nightshade, and the purple clusters hanging from the grapevine. Our acquaintance with the nauseous qualities of the one and the sweetness of the other causes us to look upon the grapes with a keener sensation of their beauty, though the berries of the nightshade surpass them in elegance of form and brilliancy of color.

Nature has distributed over her productions a certain amount of material beauty, as a seasoning to our moral enjoyments. She has thus instituted a bond between our physical sensations and our moral sentiments. Those objects which are beautiful to the eye, without the agency of thought, may be said to shine, like the sun, with unborrowed light. But other classes of objects, that possess the power of awakening pleasant emotions by suggestion, may be said to borrow their light, not like the moon from a single object, but from an infinite number of ideas and images which are reflected back upon the mind. Nature has painted the cheeks of an innocent child with the hues of the rose and the lily, that we may contemplate with the more pleasure its expressions of innocence and affection, and feel the more sympathy towards it. We follow the example of nature when we decorate the objects we love. Hence the poet, when describing a lovely heroine, paints her with the qualities of beauty, that the reader's interest may be more devoutly fixed upon her.

But the beauty of the human form and face is quite inexplicable; all our ideas of love, of delicacy, modesty, innocence, wit, vivacity, of everything it is pleasant to think of in connection with the loveliness of the sex, are awakened by the sight of a beautiful female face. It is impossible to say how much of this beauty consists in its power to excite agreeable physical sensations, and how much in its power to awaken romantic sentiments or affections by its expression of certain amiable traits of character. Yet it must be confessed that the mere beauty of color and outline in the features of the most beautiful face is but a small part of its charms.

A sense of utility and convenience is admitted by all writers to increase our estimation of a beautiful object. But this abstract consideration of utility produces only a very feeble emotion. A man seldom falls in love with a plain woman by hearing of her superior virtues as a housekeeper; but he might love her, if she was proved to possess qualities that would immediately gratify his ambition. So far as utility is concerned, the emotion of beauty is excited rather by the evident fitness of an object for conferring some passionate gratification, than by its fitness merely for serving our convenience. The sense of utility, however, enters considerably into our ideas of the beauty of landscape; and it cannot be denied that when we see an artificial decoration upon any grounds, the knowledge of its usefulness for a valuable purpose increases our feeling of its beauty.

Our moral sentiments often magnify the sensation of beauty. The pleasure we derive from beautiful objects like the rainbow is entirely independent of education, fashion, or caprice; yet this pleasure may be exalted by certain ideas acting upon the mind or the affections. A religious person, who regards the rainbow as the sacred symbol of a covenant between heaven and earth, must view it with raptures unfelt by one who has no such faith.

THE MYRTLE.

AMONG the Greeks and Romans, the oak was dedicated to Jupiter, the olive to Minerva, and the Myrtle, from the delicacy and beauty of its foliage, to Venus; and the temple of this goddess was surrounded by Myrtle groves. Hence the Myrtle and the rose have always been twined with garlands and prizes for beauty, — the one being admired for its flowers, the other for its delicate and aromatic leaves. A great deal of the romance of botany is lost to us, the inhabitants of the New World, on account of the absence from our woods of many of the plants most celebrated in classic poetry and medieval romance. We have not the heath, nor the olive, nor the ivy; and many of the humble flowers of the meadow, familiar to the reader of classical lore, are absent from our soil. Their absence, notwithstanding the beauty and elegance of many flowers and shrubs that seem to stand in the place of them, can never cease to be felt. The sacredness which a plant acquires by its association with ancient poetry and romance and with Holy Writ cannot be transferred to one of our indigenous plants of equal beauty. But there is romance in our own lives, and there are plants never mentioned in the literature of the romantic ages which are associated with certain hallowed periods and events in our youth that render them ever sacred to memory.

There are two or three plants in our own land that bear the classical name of Myrtle, not from any botanical resemblance or affinity to this plant, either in leaf or in flower, but from the aromatic odor of the leaves, like that of the true Myrtle. These plants are the Sweet-Gale, the bayberry, and the sweet-fern.

THE DUTCH MYRTLE, OR SWEET-GALE.

ALONG the low banks of rivers, and on the wooded shores of ponds and lakes that do not rise above the water-level, grows a slender and rather elegant bush, with dark and dull green foliage, possessing a very agreeable odor, which is perceived when the leaf is crushed. The Sweet-Gale is indigenous both in Europe and America. It is found only in wet places, where it forms knolls and copses, excluding all other plants by the density and vigor of its growth. This exclusive habit is owing to the multitude and tenacity of its roots, that form a subterranean network almost impenetrable. The Sweet-Gale is about half aquatic; it grows out of the water like the button-bush, and is, I believe, never found except in lands which are annually inundated.

It is this shrub that regales the sight with fresh verdure, rising out of the bosom of shallow waters in compact masses and forming little islets of shrubbery, without the mixture of any other plant. Through these wooded islets, on angling excursions, we propel our boat, while the surface of the lake is spangled with water-lilies, which, intermingled with the long blue spikes of pickerel-weed and other aquatic flowers, while the notes of the veery and the red mavis are heard from the shore, afford the scene a kind of tropical splendor.

THE BAYBERRY.

THIS species has an odor very similar to that of the sweet-gale, and from its fragrance and its waxy fruit it has obtained the name of the Candleberry Myrtle. It delights in dry pastures upon the hills and uplands, to which it is a humble, but not insignificant ornament. This plant can make no very evident pretensions to beauty, having rough and crooked branches, and imperfect flowers and fruit, without any elegance of form. But its foliage is so regular, so dense, and of so bright a verdure, that it never fails to attract attention. Indeed, it displays some of the finest masses of pure green leafage to be seen among our upland shrubbery. But seldom does any tint except the green of summer appear in the Bayberry. It takes no part in the grand pageant of autumn. The fruit of this plant is a subject of great curiosity. It consists of little greenish-gray berries, stemless, and completely covering the branches like warts, thickly coated with a waxy substance, which is soluble in boiling water. This substance, when collected, makes a very hard wax of a greenish color.

THE SWEET-FERN.

ANOTHER of those humble shrubs which, though wanting in the beauty afforded by flowers, is very generally sought and admired, is the Sweet-Fern, at the very name of which we are inspired with pleasant remembrances of spring. The Sweet-Fern is a common plant on all our hills, the close companion of the bayberry, the wild-rose, and the small kalmia. It is bound into all the nosegays gathered in May, and is a part of the garlands with which young girls crown the head of their May-queen, before the eglantine has put forth its leaves, and when the only flowers of the meadow are a few violets and anemones. This little shrub occupies a wide extent of territory, mingling its incense with almost every breeze that is scented by the rose. It is abundant in all the Northeastern States and the British Provinces.

The Sweet-Fern is a peculiar shrub, branching in such a manner as to form a perfect miniature tree, beautifully ramified with a neatly rounded head. The leaves are agreeably aromatic, and shaped unlike those of any other phenogamous plant, resembling a true fern-leaf, having alternate indentations that extend not quite to the midrib. It is a very grateful, not to say beautiful, ornament of our dry hills and pastures, and is more admired than any other equally homely shrub in our woods.

RELATION OF TREES TO THE SOIL.

I HAVE spoken of trees as the purifiers and renovators of the atmosphere, as regulators of its humidity, equalizers of the electric fluid, and as safeguards against both drought and inundations; but I have not yet alluded to the fact that they are, in dense assemblages, the actual creators, in many places, of the soil upon which they stand. The trees by means of their foliage are direct fertilizers of the ground they cover, causing it to increase in bulk as long as they stand upon it. But the leaves of trees are not the only source of this increase of bulk and fertility. The lichens and mosses, and various incrustations upon their bark, and the offal of birds, insects, and quadrupeds, all contribute to the same end. Hence the most barren situations will produce good crops for several years after the removal of their wood; and from these facts we may learn why a forest is still vigorous, though it has remained for centuries upon the same ground. If it were fertilized only by the decayed foliage of the trees, it would gradually lose its fitness to promote the health and growth of the timber. But the foreign matters I have enumerated, the decayed cryptogamous plants, and the relics and deposits of animals which have lived and died there, supply the soil with nitrogenous ingredients in which decomposed leaves are wanting.

But what are the sources of all the matters which are furnished by the trees alone? They are chiefly the atmosphere and the deeper strata of the soil. The roots of the trees, penetrating to a considerable depth, draw up from the subsoil certain nutritive salts that enter into the substance of all parts of the tree. This is restored to the surface by every tree or branch that falls and moulders upon it, and the leaves increase its bulk still more by their annual decay. According to Vaupell, "the carbonic acid given out by decaying leaves, when taken up by water, serves to dissolve the mineral constituents of the soil, and it is particularly active in disintegrating feldspar and the clay derived from its decomposition." These facts explain why the surface soil in a forest may constantly increase in bulk, without communication with any foreign sources of supply.

If a wood be situated in a valley or on a level plain, it retains all these substances for its own benefit. But if it stand upon a declivity, a part of the *débris* will be washed down by floods into the fields below. Hence, by preserving a growth of wood upon all barren slopes and elevations, the farmer derives benefit from it, both as a fertilizer and as a source of irrigation to the lower part of the slopes or the base of the hill. For some days after a rain, thousands of little rills are constantly oozing from the spongy bed of the wood, that cannot immediately become dry like an open surface. Hills, when either very barren or steep, are unprofitable alike for tillage or pasture. They require more manure than other grounds, and more labor in its distribution. Hence, if divested of wood, as I have often repeated, they are almost useless; while, if densely wooded, they fertilize and irrigate the lands below, protect them from winds, and afford a certain annual amount of fuel.

When I am journeying through the country and behold the rocky hills, sometimes for miles in extent, entirely bare of trees, and affording too little sustenance to support even a crop of whortleberry-bushes, where an acre would hardly pasture a single sheep, I am informed by the older inhabitants that these barren fields were since their childhood covered with forest. This wood cannot be restored, because the soil has been washed down from the surface into the plains below, and nothing remains to support a new growth of trees. And then I think, if our predecessors, instead of wrangling about theology, had left its mysteries to be explained by their pastors, and studied some of the plain laws of meteorology, this devastation had not taken place!

If these rising grounds, like most of the hills in New England, have a granite foundation and a comparatively barren surface soil, forests are the only means which can be used by nature to render them productive or useful in any way to the prosperity of agriculture. Were they stripped of trees, they could not long maintain their original fertility; for there is nothing to prevent the soil from washing down their sides, nothing to prevent inundations from copious rains, nothing to prevent their becoming rapidly parched by drought during a great part of every summer. Hence a mountain that is covered with a dense forest, how thin and meagre soever the soil may be from which the trees derive their support, is a source of perpetual fertilization to the lands below. Millions of living creatures, which are harbored in these woods, annually perish, leaving their remains upon the ground to fertilize it and increase its bulk. During their lifetime also, besides various substances which they have manipulated, they are constantly leaving deposits of many kinds upon the surface; and if the quantity thus spread upon a single acre of woodland could be measured, we should be astonished at the amount.

By means of forests, therefore, in favorable situations, a farmer obtains something apparently out of nothing, and makes the barren rocks and hills the sources of a part of the substances with which he fertilizes his grounds.

But I have said nothing of the pasturage afforded to cattle on the borders of woods. Out of every two or three tons of leaves which are cast upon the ground, a hundred-weight at least is but a solidification of the gases of the atmosphere. All this would be lost to the farmer, if the upper parts of his barren elevations and the sides of his steep declivities were despoiled of their wood and shrubbery. Without this forest, tons of compost produced by the annual decay of leaves would never have been created. All that proceeds from living creatures would also be lost, because they would either have never come into existence, or they would have lived and died in another place and benefited some other region.

THE VIBURNUM.

Over all the land, save where excessive cultivation and dressing of the grounds have stripped the earth of its native garniture, the roadsides are adorned with the different species of Viburnum. We detect them in winter by their many-colored branches and their finely divided spray. May clothes them with a profusion of delicate and sweet-scented flowers; lastly, autumn dyes their foliage purple and crimson, and hangs from their branches clusters of variegated fruit; so that as native ornaments of the borders of old fields and roads they are surpassed by no other shrubs. The Viburnum constitutes a great part of the underwood of our forests, thriving and bearing fruit under the deep shade of trees, but assuming a handsome shape only outside of the wood. The flowers, in circular clusters, or cymes, resemble those of the elder, but have less fragrance.

THE AMERICAN WAYFARING-TREE.

The largest and most conspicuous of this genus, and the one that seems to me to bear the most resemblance to the English Wayfaring-tree, is the Sweet Viburnum. It is a tall and wide-spreading shrub, with numerous branches and dense and elegant foliage, making a compact and well-rounded head. The leaves are single and opposite, finely serrate, and with prominent veins. Many of our shrubs produce more showy flowers, but few surpass it in the beauty of its fruit. The berries are of the size of damsons, hanging profusely from the branches like clusters of grapes. They are dark purple when ripe, with a lustre that is not seen in the grape. Just before they ripen they are crimson, and berries of this color are often blended with the ripened fruit. Like the English Wayfaring-tree, the office of this shrub seems to be to overshadow the unfrequented byways, and afford coolness and refreshment to the traveller.

THE GUELDER ROSE.

This species is common to both continents. In Europe it is cultivated under the name of Guelder Rose. In America it is known as the Snowball-tree of our gardens, and it seems to be identical with the Maple-leaved Viburnum of our woods. In the garden variety the clusters are nearly globular, consisting entirely of barren flowers, and differing from those of the wild plant in the enlargement of the florets. In the wild tree some barren florets with enlarged petals may be seen mingled with others in the cyme, chiefly encircling the disk. The fruit of this species is of a bright scarlet, and bears a superficial resemblance to cranberries, having also a similar acid taste, but a different internal structure.

THE HOBBLE-BUSH.

Why so elegant a plant as this species should bear the disagreeable name of Hobble-Bush is apparent only when we become entangled by walking over a bed of it. I have seen it frequently in Maine, where it is called Moosewood, but seldom in Massachusetts. It is never entirely erect; its principal branches spread upon the ground, while the smaller ones that bear the leaves and fruit are erect. The leaves are very large, some lobed and others heart-shaped or nearly oval. Notwithstanding its procumbent growth, it is not a homely shrub. The numerous small and erect branches that spring from the creeping boughs resemble a bed of dense low shrubbery. And when we see it in an old, dark-shaded wood, crimsoned by the tinting of autumn, and full of bright scarlet fruit, we cannot but admire it.

THE ARROW-WOOD.

Among the several species which I shall not attempt to describe, one of the most common and familiar is the Arrow-Wood, so called from the general employment of its long, straight, and slender branches by the Indians for the manufacture of their arrows. This tree seldom rises above eight or ten feet in height, and is more common in the borders of fields which are low and wet than any other species. Its fruit is of a bluish slate-color. These peculiar shrubs are often seen in the damp forest, and in the borders of wood-paths, bearing conspicuous fruit and tempting us to gather and eat, while we refrain on account of the suspicions we naturally feel when we discover the fruit of a strange plant.

AUTUMN WOODS.

WHEN the golden-rods in field and border have perceptibly faded, and we are growing weary of the monotony of summer landscape, autumn, the great limner of the forest, spreads over the earth new and enchanting pictures. Dim lights spring up daily among the shadows of the trees, and grove, copse, and thicket suffer a gradual metamorphosis. The woods are illuminated by such an array of colors that their late dark recesses appear to have the brightness of sunshine. Where a few days since there was but a shady obscurity of faded green, there gleams a luminous beauty from myriads of tinted leaves. As the twilight of the year comes on, the trees appear one after another in their new garniture, like the clouds of evening, as sunset deepens into darkness.

There is no scene in nature more purely delightful than the autumn woods when they have attained the fulness of their splendor. The sentiment of melancholy which is associated with the fall of the leaf increases our susceptibility to be affected by these parting glories of the year. So sweetly blended are the lights and colors in this gorgeous array, that no sense is wearied. The very imperfection of the hues gives a healthful zest to the spectacle, causing it never to weary like the more brilliant colors of a flower-bed. The hues of sunrise are more ethereal and exhilarating; but there is a sober mellowness in the tints of autumn that inspires the most healthful temper of mind. Far and near, from the wooded hills that display a variegated spectacle of gold, scarlet, and purple; from turrets of rocks embroidered with ferns and sumach; from old winding roads and lanes, hedged with a countless variety of gleaming shrubs, and rustic cottages half covered with scarlet creeper, down to the crimson patches of whortleberry-bushes, on the plains and in the valleys,—all is serenity and beauty.

I have often observed that the autumn woods never present that picture of gloom which is so manifest in them on a cloudy day in summer. In one respect the foliage itself is luminous, presenting warm colors that reflect light, so that the interior of a wood is actually brightened by the tinting of the leaves. I find but little pleasure in an evergreen wood at this time, unless it is illuminated by an occasional group of deciduous trees. Autumn is a sad time of the year,—the season of parting with all that was delightful in summer. The darkness of the atmosphere is even greater than in winter, when the earth is whitened by snow. We hail these warm tints of the woods, therefore, as a beneficent offering of nature for the refreshment of our spirits. All these things are beautiful even in cloudy weather, but the sun greatly enlivens the colors of the foliage, particularly when it goes down in a clear atmosphere, and every object is garnished with its beams, and mingles with golden reflections from hundreds of cottage windows. We watch their evanescent lights as they fade in the valleys and linger on the hill-tops, until twilight veils the scene in colorless shadow.

Though every one admires the beauty of autumn woods, not many are aware how imperfect are the colors that make up this gorgeous pageant. We speak of the scarlet and crimson of the maple, the oak, and the tupelo, and of many shrubs that equal them in brilliancy. But there is very little pure scarlet, crimson, or purple among these tints. If it were otherwise they would afford us less pleasure. In that case our senses would be intoxicated; now they are healthfully as well as agreeably stimulated. Pure colors spread over so wide an extent of surface would be too intense for perfect enjoyment. All the dyes of autumn foliage are sobered by the admixture of some earthy hue, something that prevents their rivalling the tints of heaven.

Green and yellow are often seen in their purity in the leaves of trees; crimson and scarlet are seldom pure, except in some parts of the brightest leaves. Even their green is not perfect, save in that stage of their development that precedes their full expansion. After this period, as the landscape-painter well knows, all verdure is tarnished and rusty. Indeed, the colors of leaves will not bear comparison with those of flowers, either in purity or variety; yet when viewed from a distance, and illuminated by sunshine, they seem nearly pure. Red leaves of different shades in sunshine produce at a distance the effect of crimson or scarlet, chocolate hues that of purple, and browns that of orange.

The hues of autumn are not very conspicuous before the middle of September, and it is worthy of notice that the brightest and purest colors are seen at the time when three fourths of the trees still remain unchanged. As one after another assumes its ruddy, golden and purple hues, the earlier and more brilliant drop their leaves; and some are entirely denuded, while others are fully covered with foliage and verdure. Even different individuals of the same species, of maples especially, manifest a great difference of habit in this respect, caused in some cases by the peculiarities of their situation. Trees in swamps and low grounds lose their leaves earlier than the occupants of a deep soil in the uplands.

Some species are perfectly uniform in their colors. The poplar and birch, for example, are invariably yellow;

the sumach and whortleberry are chiefly red; while the maples display as many colors as if they were of different species. But each individual tree shows nearly the same every year, as apple-trees bear fruit of the same tints from year to year. Two red maples growing side by side are seldom alike, and in a group of them you will see almost as many shades of color as trees. Some are entirely yellow, others scarlet, some crimson, purple, or orange, others variegated with several of these colors. There is more uniformity in the tints of the sugar-maple. I have seen long rows of this species that were only yellow and orange, though its colors generally vary from orange to scarlet. Purple and crimson are confined chiefly to the red maple; I have seen in different individuals of this species all the hues that are ever displayed in the autumn woods. The red maples, more than all other trees combined, are the crowning glory of a New England autumn. The sugar-maple, though more brilliant, has a narrower range of colors.

As early as the last week in August, we perceive the tinting of a few red maples, which always exhibit the earliest change. Sometimes a solitary branch is tinted, while the remainder of the foliage is green, as if something affecting its vitality had prematurely colored it. Frequently the coloring process begins at the top; the purple crown of autumn is placed upon the green brow of summer, and we behold the two seasons represented at once in the same tree.

The first coloration is usually seen at the veins of the leaf, extending outwardly until the whole is tinted. Sometimes it appears in spots, like drops of blood upon the green surface; and in this case the leaf usually remains spotted. In the foliage of trees that assume a variety of colors, yellows generally predominate in the interior of the mass, red and purple on the outside. In the red maple, and less frequently in the rock-maple when in a protected situation, the leaves are often formally variegated with figures of yellow, red, green, and purple. Those of the poison-sumach, the cornel, and the snowy mespilus, are sometimes beautifully striated with yellow or orange upon a darker ground; but I have searched the woods in vain to find any other than a maple-leaf configurated like a butterfly's wing.

In the foliage of the tupelo deep shades of purple first appear, brightening into crimson or scarlet before it falls. This tree more invariably shows a mass of unmixed crimson than any other species. Even in the maple, if the general presentation is red, you will find a considerable mixture of yellow. The colors of the scarlet oak are seldom pure or unmixed; but those of the tupelo are invariable, except as they pass through the gradations from purple to scarlet. If, therefore, the tupelo were as common in the woods as the maple, it would contribute more splendor to the scenery of autumn. There are many trees that never produce a red leaf. I have never found one in the foliage of the poplar, the birch, the tulip, the hickory, or the chestnut, which are all of some shade of yellow; but there are usually a few yellow leaves scattered among the ruddy foliage of any tree that assumes this color.

When all the circumstances attending the season have been favorable to the tints of autumn, there is no tree of the forest that would attract more admiration from the beautiful sobriety of its colors than the American ash. But this tree is so easily affected by drought, that after a dry summer its leaves fall prematurely and its tints are imperfect. The colors of the ash are quite unique, and distinguish it from all other trees. Under favorable circumstances its coloring process is nearly uniform. It begins with a general impurpling of the whole mass of foliage nearly at the same time, and its gradual changes remind me of those observed in sea-mosses during the process of bleaching. There is an invariable succession in these tints, as in the brightening beams of morn. They are first of a dark bronze, turning from this to a chocolate, then to a violet brown, and finally to a salmon-color, or yellow with a slight shade of lilac. When the leaves are faded nearly yellow, they are ready to drop from the tree. It is remarkable, that, with all this variety of hues, neither crimson nor any shade of scarlet is ever seen in the ash. It ought to be remembered that the gradations of autumn tints in all cases are in the order of those of sunrise, from dark to lighter hues, and never the reverse. I make no reference to the browns of dead leaves, which are darker than yellow or orange, from which they turn. I speak only of the changes of leaves before they are seared or dry

After the middle of October, the oaks are the most conspicuous ornaments of the forest; but they are seldom brilliant. In their foliage there is a predominance of what we call leather-colors, with a considerable mixture of certain shades of red that are peculiar to the oak. We rarely find pure yellow or scarlet leaves in the foliage of any species of oak. The color of the scarlet oak is nearer a purple or crimson than any other shade of red. The white oak turns, with but little variation, to an ashen-purple or impure violet. The black and red oaks display varying and imperfect shades of drab and orange. The oaks are remarkable for the persistence of their foliage, and for the duration of their tints, which are chiefly the brown and russet of dead leaves with a lively polish. Long after other deciduous trees have become leafless, the various sombre shades of the different oaks cast a melancholy tinge over the waning beauty of the forest.

We are wont to speak of trees as the principal objects

of admiration in autumnal scenery, but the shrubs, though less conspicuous on account of their inferior size, are not less brilliant. It is also remarkable that reds predominate in the shrubbery, and yellows in the trees. Reds and purples distinguish the whortleberry, the cornel, the viburnum, and the sumach, including all their species. There is indeed so small a proportion of yellow in the shrubbery, that it is hardly distinguishable in the general mass of scarlet, crimson, and purple. Among trees, on the contrary, yellows prevail in all miscellaneous woods. They distinguish the poplar, the birch, the hickory, the tulip-tree, the elm, and a good proportion of the maples. It ought to be remarked, however, that there are more shrubs than trees that do not change materially, but remain green until the fall of their leaves. The alder remains green; and as it covers a large proportion of our wet grounds, it might seem to an observer in those situations that the tints of autumn were confined to the trees.

Many persons still believe frost to be the great limner of the foliage, as if it were a sort of dyeing material. On the contrary, the slightest frost will destroy the tints of every leaf that is touched by it. It is not uncommon to witness a general tarnishing of the autumnal tints by frost as early as September. In some years they are spoiled by it before they have begun to be developed. An autumn rarely passes when the colors of the foliage are not half ruined before the time when they ought to be in their brightest condition. But the injury they receive from slight frosts is not apparent to careless observation. In the meridian of their beauty, heat will damage the tints as badly as frost. A very hot and sunny day occurring the first or second week of October makes almost as much havoc with the ash and the maple as a freezing night, fading their leaves rapidly and loosening their attachment to the branches, so that the slightest wind will scatter them to the ground. Yet the action of heat differs materially from that of frost. Frost imbrowns and crisps or sears the leaves, while heat only fades them to lighter and more indefinite shades. Frost is destructive of their colors, heat is only a bleaching agent. Cool weather in autumn without frost is necessary for the preservation of its seasonal beauty.

The most brilliant autumnal hues appear after a wet summer, followed by a cool autumn, unattended with frost. Cool weather preserves not only the purity of the colors, but also the persistence of the foliage. If the early frosts are delayed, the tints are brighter for this delay while the weather remains cool. But a wet summer is so generally followed by premature cold, that the finest displays of autumn scenery are often suddenly ruined by a hard frost. Seldom are all the favorable circumstances for preserving the purity of the tints combined in any one season. Not more than once in six or eight years are both heat and frost kept away so as to permit the leaves to pass, unseared and untarnished, through all their beautiful gradations of color.

There are several herbaceous plants that display tints similar to those of the woods; but they are not very conspicuous. I must not fail to mention the samphire, a plant of the salt marshes, possessing no beauty of form, having neither leaves nor any very discernible flowers, which every year contributes more beauty of color to the grounds it occupies than any flower of summer. Though I have seen no printed account of its magnificent crimson spread interruptedly over miles of salt marsh, my attention has often been called to it by ladies, who are more sensitive than the other sex to such appearances, and more careful observers of them.

The tints of the forest in America are said greatly to surpass those of the European woods. Having never visited Europe, I cannot speak of the comparison from my own observation. But from descriptions of them by different authors who have treated the subject, I have been led to believe that the difference is caused by a larger admixture of scarlet and crimson among the tints of our own trees. To aid the reader in drawing a comparison between them, I have made a synopsis of the tints of American woods during September and October; and have copied a similar one, less full and particular, by George Barnard, of English woods.

NOTE.

There are a few trees and shrubs, of which the alder and buckthorn are examples, that so seldom show any kind of a tint that I have not included them in my list; and there are several species of oak that display such a motley combination of green and rust, with faint shades of purple and yellow, that it is impossible to classify them. In my list I have only named the genera, except when the species are distinguished by important differences. The brown hues of the oak and the beech are the tints only of their dead leaves or dead parts of leaves; but pure browns are sometimes seen in the living leaf of the snowy mespilus, the pear-tree, and the smoke-tree; in others they occur so seldom that they may be classed as accidental hues. I ought to add that only a small part of what may be said of the tints of trees is unqualifiedly correct. They are greatly modified by circumstances which cannot always be understood. I have seen maples that always remained green, apple-trees dressed in scarlet and yellow, and lilacs in a deep violet; but I have never seen a purple, crimson, or scarlet leaf on any of the trees of Division I. of the Synopsis.

SYNOPSIS OF THE TINTS OF DIFFERENT TREES AND SHRUBS IN SEPTEMBER AND OCTOBER.

Division I.

Trees and Shrubs that display Yellow Tints alone, without ever a Purple or Red Leaf of any Shade.

Althæa. Bayberry. Clethra. Dutch Myrtle. Elder. Locust. Privet. Willow.	Section 1. — Verdure of summer unchanged, or with a slight and sometimes a considerable mixture of yellow leaves, before they fall.
Black Walnut. Butternut. Catalpa. Chestnut. Elm. Grapevine. Hickory. Horse-Chestnut. Lime. Plane. White Birch.	Section 2. — A general mixture of rusty green and yellow, sometimes pure yellow under favorable circumstances. The rust attaches only to dead leaves or to the dead parts of leaves.
Beech. Birch. Honey Locust. Mulberry. Poplar. Tulip-tree. Green-Brier.	Section 3. — Pure yellow, of different shades.

Division II.

Trees and Shrubs that display all Shades of Purple, Red, and Yellow.

Apple-tree. Barberry. Blackberry. Cherry. Hawthorn. Lilac. Missouri Currant. Mountain Ash. Pear-tree. Peach-tree. Plum-tree. Quince-tree. Raspberry. River Maple. Rose. Spiræa.	Section 1.—A predominance of green, with a slight and sometimes a considerable mixture of purple, red, and yellow, of different shades. All the rosaceous plants are included in this section or the following. Individuals of some of these species are occasionally brilliant.
Blueberry. Cornel. Hazel. Poison-Ivy. Scarlet Oak. Smooth Sumach. Strawberry-tree. Tupelo. Velvet Sumach. Viburnum. Virginia Creeper. White Oak. Whortleberry.	Section 2.—Purple, crimson, and scarlet, with only a small mixture of yellow, if any.
Mountain Maple. Red Maple. Rock Maple. Poison-Dogwood. Smoke-tree. Snowy Mespilus. Striped Maple.	Section 3.—Variegated tints, comprising all shades of purple, crimson, scarlet, orange, and yellow on the same tree, or on different trees of the same species. Leaves often striated, and sometimes figured like a butterfly's wing.
The Ash.	Passing through all shades from a dark chocolate to violet, brown, and salmon. The ash is perfectly unique in its tints, having no reds, and being the only tree that shows a clear brown as one of its regular series of tints in the living leaf.

[From George Barnard's "Drawings from Nature."]

Calendar of the different Tints assumed by various Trees toward the End of September (in England).

English Maple. The leaves of the maple change first of all to an ochrey yellow, then to a deeper tone.

Ash. Fine lemon yellow, soon falling and leaving bunches of seeds of a brown hue.

Hornbeam. Bright yellow.

Elm. Generally orange, but with some irregular patches of bright yellow

Hawthorn. Tawny yellow, but greatly modified by tones of deep reddish-brown, and brilliant clusters of berries.

Hazel. Pale ochrey yellow, with browner shades for the clusters of nuts.

Sycamore. A dull brown.

Oak. Yellowish green.

Horse-Chestnut. A great variety of beautiful rich hues, from a pale yellow to a bright crimson orange [?].

Beech. Also finely varied in color, but more of a maroon color than the chestnut.

Cherry. Most diversified and charming, in tints of yellow, red, crimson, maroon, and purple.

NOTE. — I perceive that the author does not distinguish between the tints of living and seared or dead leaves.

THE CORNEL.

The different species of Cornel abound in all places occupied by the viburnum, to which they bear a superficial resemblance, though the two genera are not allied. They are graceful and rather prim-looking shrubs, having a hard and close-grained wood, and containing in their bark a large proportion of the bitter principle of the cinchona. Their leaves and branches are opposite, which increases their resemblance to the viburnum. They are very abundant in the Northern States; and it is remarkable that the different species might be distinguished by the colors of their fruit. The Florida Cornel, called the Flowering Dogwood, bears scarlet berries; there is also a purple-fruited Cornel, a white-fruited and two blue-fruited species, one leaden-colored, and in Canada a species with dark brown berries.

It is seldom that the species of any genus of plants differ in the opposite or alternate characters of the leaves and branches. But the purple-fruited Cornel is called alternate-leaved, to distinguish it from the other species. It is not, however, a genuine exception; for the leaves come out around the stem, not in a true alternate arrangement, but in imperfect whorls, and mixed with some that are opposite. The flowers are small, in irregular cymes; the fruit of a dark purple. It is found in swamps and low moist woods, and, with the other species, constitutes a fair proportion of the underwood of our deciduous forests.

The white-fruited Cornel is very frequent by waysides,

rising a little above our loose stone-walls. This seems to be the most abundant species outside of the woods in the vicinity of Boston. Its flowers are white and rather inconspicuous, and are succeeded by clusters of pearly white berries. The blue-fruited Cornel, or red osier, is remarkable for its colored branches and large round leaves with an acuminate termination. The blue fruit of this species is very ornamental, and it is distinguished after the fall of the leaf by its bright red stems and branches. The Cornel is hardly less important than the viburnum in adding variety to our wood-scenery at all seasons.

By far the most interesting and beautiful species of the genus is the Florida Cornel, so called from its abundance in the forests on the American side of the Gulf of Mexico. In all that region, the woods in May are white with its large conspicuous flowers, sometimes occupying tracts of many acres exclusively, covering them with an almost unvaried whiteness, before the leaves of the trees are put forth. The flowers are borne in semi-globular heads, enclosed in a large spreading involucre, which is often mistaken for the corolla, the florets within resembling superficially a collection of stamens. About the first of June, in New England, these trees are very attractive, seeming like masses of pure white inflorescence. In the North it does not constitute the principal growth of any wood; but it is admired by all when they see it scattered among the greenery, and admired the more from its infrequency in this region.

The small branches are greenish, striated with longitudinal and irregular white lines. The leaves are two or three inches long, oval, and of middle size. The flowers appear on the ends of the branches, included in an involucre consisting of four divisions. The head of florets thus enclosed ripens into a bunch of bright scarlet berries, surrounded by a dark purple calyx. In the autumn all the species turn to different shades of red and purple.

The little dwarf Cornel, though an herbaceous plant, deserves mention in connection with the other species. It may be compared to a flower cut off with a single whorl of leaves, and then inserted into the ground. You might suppose that the large tree Cornel was buried, and that these little whorls, with their flowers, were peeping up through the ground from the branches beneath. At some distance they are easily mistaken for wood-anemones, though on examination no resemblance is apparent. The flowers are very showy and attractive in the wild pastures and woods, and produce in the autumn a round and compact cluster of scarlet berries, which are said to be pleasant and wholesome, but rather insipid. In winter they are the food of many species of birds.

MOUNTAINS.

MOUNTAIN scenery has always been admired, and will never cease to charm the eyes and excite the imagination even of the dullest of mankind. However reluctant we might feel to be surrounded by contiguous mountains, and imprisoned in their valleys, we are all delighted with a journey that leads us through their romantic passes and over their fearful heights. Every stage in our progress opens a new scene to our eyes, amusing us alternately with confined and extensive views, on the outside of a range often sublime, and affecting the mind with a singular exhilaration. Great altitude is one of the most remarkable sources of sublimity arising from position; and the emotions produced by it are the more vivid, when we have just emerged from some green pastoral valley.

Mountains, except on the outside of a range, are unfavorable to extensive prospects, and the sublimity of this kind of scenery cannot be felt when passing along through their valleys. Prospects of the grandest description are frequent; but the inhabitants are for the most part shut out from all chance to look abroad upon the earth, or even to see the rising and the setting sun. The distant view of a mountain rising into the clouds, and enveloped in a misty obscurity that enhances our conception of its magnitude, is always attended with an emotion of grandeur, very similar to what is felt on viewing the surrounding landscape from its higher elevations. A sense of sublimity may thus be excited in imaginative minds by contemplating a mountain from below; but in general a prisoned sensation must be felt in its valleys, from the conscious restraint upon our freedom. Here we have no breadth of prospect, as we should have from an island, and feel more as if we were confined by walls, though we might emerge from it more easily than we could escape from the island.

The moral influence of a permanent home in a mountain valley is highly favorable to cheerfulness, by increasing our susceptibility to be agreeably affected by the scenes that are spread out above and beyond us, which can be seen only by leaving the valley. Our egress from this retreat is always exhilarating and hopeful, because our journey is upward, and every step widens the circumference of our horizon. But if our home is on an elevated site on the mountain, that affords us an apparently boundless prospect, its influence must be depressing, because we cannot enlarge our prospect by leaving our situation. Our journey into the world is downward. Every step narrows our landscape, brings objects that were grand and beautiful in the distance so near to us as to be tame and uninteresting. I can believe, therefore, that, if a man were subject to melancholy, he would find his cure more certainly by making his home in a valley than upon a mountain.

For a permanent residence I should prefer a plain, with a view of mountains afar off, to a valley among them, or to a mountain. The exhilaration produced by a wide prospect and great altitude is a tone of the mind that cannot be long sustained, and we should lose our susceptibility to be affected by it if it were constantly before our sight. But a view of distant mountains is not exhilarating. It acts quietly upon the imagination, when the mind is in a reflective mood; it is never glaring, and affords no unnatural stimulus. The same may be said of any remarkable prospect from our doors and windows, if placed conspicuously before us. We gradually lose our power of enjoying this and similar views. Hence our daily and familiar prospects should not be of a stimulating character; for everything that exhilarates, when habitually used, deadens the sensibility. The terrestrial views about our home should be quiet and interesting, but not extraordinary, to preserve a healthy tone of the mind, as our daily food should be plain and simple to preserve the health of the body.

The same remarks do not apply to celestial views. The sky, during a great part of the day, is a mere canopy of light. Its exciting effects are felt only on extraordinary occasions, which are transient. It is beautiful when the sun rises and when it sets, enveloped in highly refractive vapors, and sublime when curtained by illuminated masses of finely organized clouds. But these spectacles are not liable to tire us by their frequency or duration. Give me, therefore, a clear and unobstructed view of the heavens, from my place of constant residence, that I may witness those momentary spectacles of beauty that occur in the morning and evening.

It is a popular error to suppose that the inhabitants of mountains acquire from their habitual prospects a lively imagination and an expanded mind. The influence of the scenes around them would be quite the contrary, if it were felt at all, since they are mostly confined in valleys, and shut out from the world. Yet it is not their limited view of the heavens and the earth that would narrow their minds, but their want of intercourse with men; for mountaineers are seldom engaged in commerce, the grand enlightener of nations. They are herdsmen and tillers of the soil, and by living apart from other men they acquire a clannish spirit and become addicted to superstition and fanaticism. It is also believed that the inhabitants of mountains are more warlike than the inhabitants of plains. Their superior success arises from the greater facilities afforded by their position for practising the artifices of war. They know all the passes and all the grand points for intrenchment and attack. This knowledge and these advantages have gained them an undue reputation for courage and heroism.

Nature has confined her intellectual and moral gifts to the inhabitants of no particular description of surface. A constant familiarity with the sublime scenes of nature does not exalt the imagination; neither does a confined valley, with its narrow prospects both of the heavens and the earth, cramp the mind or dull the sensibility. It is the want of education and of intercourse with the world that affects the character. It cannot be denied that there is a moral influence arising from landscape; but, contrary to the general opinion, tame and rather homely landscape is the most favorable. All those scenes that enchant with their beauty, or dazzle and intoxicate by their grandeur, when constantly before us, are depressing, and cause the same action upon the mind that narcotic stimulants produce upon the nervous system. But the majority of inhabitants of every country are unaffected in any way by their daily and habitual prospects. Their exhilarating and depressing effects are chiefly felt by persons of more than ordinary culture or poetic sensibility.

It is to individuals of this class, however, that my remarks are addressed. They will agree with me that a moderate share of the beauty and sublimity of landscape, in the scenes about our home, is the most desirable, on account of their moral influence. Homely objects and monotonous views are not depressing; they are simply unsatisfying; and if we are within the reach of fine prospects, we are always prepared to enjoy them. Landscapes, therefore, except within very narrow limits, ought never to be highly dressed or decorated. We should leave all such ornamentation to Nature, who, while she provides endless scenes of beauty for those who seek them, never clogs our sight by their profusion. Though the influence of moderately pleasant natural scenery is healthful and never tiresome, I can imagine nothing so melancholy and depressing as a country universally dressed in the highest style of English landscape art.

My object in these essays is to present the reader with pictures of those scenes which are common, and that fail to attract attention only because the generality of men can see nothing admirable in Nature except her monstrosities. I am not obliged to visit Mount Washington or the Falls of Niagara to experience the effect of that sublimity which I can equally perceive in the fading fires of the heavens at sunset, or in their starry glow by night. The common scenes of Nature are capable of affording the most intense delight to those who have accustomed their minds to the study of all her aspects. We may sail round the globe in quest of scenes of grandeur and beauty; but we shall seek in vain for anything more beautiful than the rainbow, or more sublime than the sun emerging, as it were, from the ocean at sunrise, enshrouded in the dappled hues of morning.

THE SUMACH.

THE Sumachs are not the objects of any special admiration. They are not the favorites of nature or of art, neither adding dignity to the landscape nor expression to the canvas of the painter. But they blend their fine pinnate foliage with the wayside shrubbery, varying its appearance by their original habit of growth; and they are seen springing in little groups upon sandy plains, where they relieve the eye that might otherwise be wearied with the monotonous waste of sorrel and tufted andropogons. They display many of the characters of the tropical plants in their long compound leaves, and in the exuberant growth of their recent branches. They are distinguished by their milky, resinous, and in some cases poisonous sap.

THE VELVET SUMACH.

The most common and conspicuous species in New England is the Staghorn, or Velvet Sumach, the largest of the genus. Its name is derived from a certain likeness of its crooked branches, when deprived of their leaves, to a stag's horn. This Sumach rises to the dignity of a tree in favorable situations, and soon becomes a handsome standard, if the suckers about the roots have no chance to grow. Though its branches are crooked and irregular, and form a spray that is absolutely ugly, the tree is very comely when wearing its leafy garniture and decked with conical bunches of crimson fruit.

The Sumach is sometimes very ornamental in situations that permit the whole ground to be occupied by it. Its natural habit of growth is in clumps, gradually spreading over a wide extent of surface. So prone is this tree to throw up suckers from its long roots, that if it meets with no opposition it is apt to monopolize the whole ground. The most appropriate places for it are the banks of railroads and other similar slopes, which are rendered firm by the network of its numerous roots. There is no other plant that would in so short a time cover a gravelly bank with wood and foliage.

The Smooth Sumach is a smaller shrub, averaging only three or four feet in height. It affects similar localities, being common on the borders of dry fields and the sides of old roads that pass over a sandy and gravelly plain. It is not readily distinguished from the larger species; but its fruit and flowers are borne in loose panicles, and its bunches have none of that downy substance that characterizes the Velvet Sumach.

THE POISON SUMACH, OR DOGWOOD.

I come now to speak of the Bohon Upas of our land, — the Poison Sumach. This is confessedly a dangerous plant, and is allied to the shrub from which the celebrated Wourali poison is made by the natives of Guiana. The poisonous properties of the sap are said to be dissipated by boiling. Hence the varnish prepared by the Chinese from the sap of this plant is free from its injurious properties. Hence also the danger of being exposed to its fumes, when its branches are burned with other brush.

The Poison Sumach is a very elegant shrub. It is prim and slender, and draws attention by its want of resemblance to other trees and shrubs in our woods. The main stems and principal branches are of an ashen-gray color, though the recent shoots, before they harden into wood, and the leaf-stems are of a fine crimson or purple. The leaves are beautifully pinnate, of a light green hue with purple veins. The flowers and fruit are greenish, inconspicuous, and without any beauty. This plant, unlike the other species, is found only in low boggy situations.

There are some unaccountable facts connected with the poisonous qualities of this tree. While some persons are affected with dangerous swellings and inflammation on the least exposure to it, others handle it, breathe its burning fumes, and even chew its leaves and branches with impunity. Some are rendered more susceptible by having been once poisoned; others, who were often injuriously affected by it in their youth, outgrow their susceptibility, and may afterwards handle the plant without danger. As certain persons are exempt from the malignant effects of this plant, there is occasionally an instance of similar effects suffered by individuals from other plants. I am acquainted with a lady who has been frequently poisoned by handling the branches of the black wild-cherry. Such isolated facts serve to increase the mystery attending the subject.

A notion prevails in the country, that the recent shoots of the pitch-pine, if frequently chewed, will render any one safe from the effects of this poison. The forest undoubtedly abounds in antidotes to the injurious action of the Poison Sumach and other similar plants; and I have often thought that the impunity with which the goat browses upon narcotic herbs may be caused by the counteracting effects of other plants among the many species which he devours in the field and pasture. It is admitted that persons who spend much of their time in the woods are not liable to be affected by this poison. They may, in some way or other, become inoculated with its antidotes. I have never suffered in the least degree from it, though I have passed a considerable part of my lifetime in the forest. Catesby mentions a fact, which he says was well attested, of an Indian who daubed himself with the juice of the purple bindweed, and then handled a rattlesnake with his naked hands with impunity. Some high authority may be quoted to sustain any similar improbable fact or absurd opinion.

THE ELDER.

Everybody is familiar with the Elder, with its large corymbs of white flowers, hanging over ditches and water-courses, rivalling the linden in sweetness and equalling the balm in its healing virtues. It is common in all wet fallows, flowering in the latter part of June. No shrub is so generally known, both as a tenant of the fields and as an ingredient in the packages of the simpler. We have seen its dried flowers in nice paper bags, neatly done up by some benevolent hands for the benefit of the sick, and we breathed their odors as they were wafted from the vessel in which they were steeped, before we ever saw them in the fields. The Elder is one of the flowering shrubs that first attracts our attention after the blossoms of the orchard have faded. The bee is seen to hunt for it before the vine is in blossom, leaving the flowers of the garden for these abundant stores of native sweets. In autumn we have seen the fences and brook-sides laden with its fruit, while the purple clusters were stripped day after day by the robin and catbird, until not one was left to fall to the ground. When the leaves are gone, the branches are sought by children, who use its hollow wood for making various juvenile implements.

"The Elder," says Barnard, speaking of the English plant, "is common, almost universal, in cottage gardens, hedge-rows, and ruins. It is in fact a thoroughly domesticated tree, and seldom is it found in England far from human habitation, although I have seen it in the wildest valleys of the Pyrenees, when it appeared to have the richest scarlet berries, instead of black." The species seen in the Pyrenees is probably identical with the American panicled Elder, a rare species in New England, bearing its flowers in spikes, and producing scarlet berries.

The Elder has not much beauty when unadorned either with flowers or fruit. Its pinnate leaves are of a dull green, and seldom add any tints to the glory of autumn. Its flowers, borne in large flat cymes, are very showy, and emit a peculiar though agreeable odor, and are used in Europe to give to wine the flavor of Frontignac. The berries of the European Elder, which is believed by Michaux to be the same as the American common Elder, differing only in its superior size, are said to be poisonous to poultry. But the fruit of the American shrub possesses no such properties. It is eagerly devoured by the insectivorous birds, and is used in the manufacture of a harmless dietetic wine, whose benefits have been very generally appreciated by nostrum venders.

RUDENESS AND SIMPLICITY.

WHEN making a pedestrian journey I always follow the rudest paths, unless the whole surface is wild and rugged. Especially in the suburbs of cities, I avoid the roads that would lead me past elegant villas, and turn aside into the more homely parts of the town, where I may behold some pleasant reminders of simple and humble modes of life. I am willing to see an occasional dilapidated house, with its broken fences, its burdocks, and its neglected garden. These hardly lessen the pleasure afforded me by the sight of neat little cottages, in many different styles, their modest gardens and flower-beds, their playful children around their doors, unencumbered with elegant restraints, and the general signs of unambitious comfort. I prefer these neat and homely scenes to the most beautiful prospects when marred by ostentatious decorations.

A great deal has been said and written in these days by lecturers and essayists to persuade men to love and appreciate the "beautiful." But the multitude are very far from needing any such lessons; as well might we waste words in persuading them to seek pleasure. It is rude landscape and homely objects, having a charm about them that is more admirable than beauty, which men in general can learn only by tuition to appreciate. All men have an innate love of the "beautiful," — a native passion for paint, feathers, and brocade. To children and boors a border of peonies is vastly more attractive than the rude scenery of our New England hills; and the love of flashy colors manifested by barbarians has always been remarked.

Rude people are not admirers of homely landscape. They are unaffected by scenery, unless it is either grand or elegant. But the man of educated mind — the painter or the poet — sees a charm in rudeness surpassing the most admired scenes which are merely beautiful. He prefers Nature in her own spontaneous dress. He loves that simplicity which would be spoiled by any attempt to improve or embellish it.

As we advance from childhood to maturity, and grow in knowledge and culture, the scenes of nature become full of associations, some pastoral or romantic, some grand or sublime, others rude, weird, or desolate. The different sentiments awakened by them enter into our ideas of the picturesque, from which that of mere elegance is generally excluded. That there is a specific charm in rudeness is acknowledged by all writers on art, and some have even regarded it as essential to the picturesque. But they have generally believed it confined to pictures of scenes and objects, and that the sentiment cannot be awakened by the things represented. A picture of an old building, for example, they would say, is admired when the building itself would attract no attention and excite no emotion. It may be said, however, that there are more who can understand the expression of a picturesque object in a painting than in reality, because the imagination of the spectator is guided and assisted by the painter in one case, and not in the other. But the painter, or any one who views nature with a painter's eye, would be as much affected by the natural scene as by the picture that represents it.

I am persuaded that a highly agreeable and specific emotion is awakened by the sight of rudeness in landscape, that deserves a place in the category of agreeable sensations derived from nature. Rudeness of scenery corresponds with simplicity of life. It is the opposite of elegance, under many circumstances not less charming, and always more picturesque. Elegance is suggestive of art and painstaking; rudeness, of freedom and nature. Elegance reminds you of effort; rudeness, of repose. The one is associated with freedom, the other with exclusiveness. We see the fondness of people for the sensations produced by each of these qualities displayed in rude flower-baskets, made of neatly joined pales, with the bark side outward, and filled with flowers elegant both in colors and arrangement. But of the two qualities elegance will sooner weary the mind by demanding more attention and affording more stimulus to the sight. In like manner, at our tables, luxurious viands sooner pall upon the appetite than those which are plain and simple. But while the opposite of elegance is agreeable in a natural scene, this cannot be said of the opposite of neatness. A cantonment of Irish laborers calls up none of those images of combined neatness and rusticity which are awakened by the scenes of a genuine New England hamlet.

I have often seen in the country, a few paces back from the road, certain plain cottages, which are not described in books of taste and art, that seem to me perfect examples of neatness and simplicity, without any approach to elegance. The rudeness associated with them is in the grounds and fences, more than in the buildings. Such houses are now more common in Maine than in Massachusetts. One of these rustic buildings in particular has often attracted my attention. It has never been painted, and the dark stone color of its walls is in pleasant consonance with the green in front and on each side of the house. This is kept closely fed by the farmer's cows, which are allowed to graze upon it after returning from pasture. No fence encloses this beautiful plat of verdure, which is shaded by two drooping elms. Beneath one of them is a well, with a plain, unadorned curb, and in the rear of the house a quantity of wood is neatly piled against the rugged stone-wall. There is an absence of all litter about the house; and the superfluous branches which have been lopped from the orchard trees are cut up for fuel, and thrown into a conical heap a few steps from the back door. A foot-path winds along the roadside to the front, and another in the rear of the house leads to the field or garden. There is neither paint nor white-wash anywhere to be seen, yet all who pass by would point to the place as a pattern of neatness.

But Fashion, the idol of vulgar minds, not content with exercising her sway in the palatial mansion and at the haberdasher's shop, in deciding the comparative merits of insipid pears at the horticultural rooms, and displaying her metaphysical subtlety in wine-cellars, became, a few years since, a teacher of æsthetics. Since that time "men of taste" have been at work, in book, pamphlet, and lecture, casting contempt upon rudeness and simplicity, always reserving a little praise for their counterfeit. According to their ideas, nothing in village scenery should be tolerated that is not "beautiful." They would destroy or remove out of sight every object that does not bear on its face the gilding of wealth or the flummery of art. Already, in many of our villages once charming in their simplicity, do we see the effects of this esthetic vandalism. How many a delightful place, which we could not look upon without imagining it a little nook in paradise, has been destroyed by the prevailing taste for beautifying one's abode! How many comfortable old farm-houses, with their neat and rustic enclosures, their tussocks of wild shrubbery that afforded a harbor to the birds, and their pleasant approaches by a foot-path fringed with wild flowers, have been improved and beautified, until we turn from them with mingled disgust and contempt, as from an aged crone dressed in jewels and laces!

But why is this expression of rudeness, when combined with other agreeable qualities of landscape, so charming? Why do we prefer a country that is marked with a certain degree of wildness to one that is highly ornate and beautiful? Because we love the evidences of a happy state of society. We would not that nature should remain a wilderness, nor be transformed into a garden, because the one indicates a savage state, the other a degree of luxury that is incompatible with the best welfare of man. We would behold in our rural prospects the traces of a hardy and virtuous population. If Nature herself has become effeminate, what can we expect of her children? If Nature be dressed like a courtesan, will the children of the swains who live upon her bounties be contented with the humble emblems of industry, — the reaping-hook and the wheaten sheaf? I prefer those appearances which are tokens of virtuous frugality, temperance, and industry, to the most admired scenes of grandeur and luxury. For this cause do the moss-grown rocks by the hillside, fields of grain and orchards surrounded by the rude landscape of nature, plain farmhouses smiling amid the golden products of independent labor, affect us with delightful sentiments, as evidence of those healthful habits which are not yet banished from the land by luxurious improvement.

THE HEATH.

THERE are no heaths in New England, or on the American Continent. We know them only as they are described in books, or as they are displayed in greenhouses. We are strangers to those immense assemblages that furnish an uninterrupted vegetable covering to the earth's surface, from the plains of Germany to Lapland on the north, and to the Ural Mountains on the east. These plains, called heaths or heathlands, are a kind of sandy bogs, which are favorable to the growth of the Heath, while other plants with these disadvantages of soil cannot compete with them. The tenacity with which they maintain their ground renders them a great obstacle to agricultural improvement. They overspread large districts to the almost entire exclusion of other vegetation, rendering the lands unfit to be pastured, and useless for any purpose except to furnish bees with an ample repast but an inferior honey.

It is often lamented by the lovers of nature that the Heath, the poetical favorite of the people, the humble flower of solitude, the friend of the bird and the bee, affording them a bower of foliage and a garden of sweets, and furnishing a bulwark to larks and nightingales against the progress of agriculture, — it is often lamented that this plant should be unknown as an indigenous inhabitant of the New World. But if its absence be a cause for regret to those who have learned to admire it as the poetic symbol of melancholy, and as a beautiful ornament of the wilds, the husbandman may rejoice in its absence. We have in America the whortleberry, whose numerous species and varieties occupy, like the heaths of Europe, those lands which have not been reduced to tillage, without depriving them of their usefulness to man. They become in their beneficent products a source of profit to thousands of indigent gleaners of the pastures, and of simple luxury to all our inhabitants. Though Nature has denied us the barren flower, and left the imagination unrequited, she has given us, in the place of it, a simple fruit that furnishes annual occasions for many a delightful excursion to the youths and children of our land, and is a simple blessing to the poor.

The farmers of Eastern Massachusetts, who have seen the dyer's broom spread itself over the hills, occupying the whole ground, and entirely displacing all valuable herbs and grasses, may form some idea of the mischiefs attending the spread of the Heath in Europe. The heaths might be described as tree-mosses, bearing a multitude of minute campanulate flowers of various colors. They are not exceeded by any other plants, except mosses, in the uniform delicacy of their structure. Hence they are admired by florists, who find among them those multitudinous varieties which, in other plants, are produced by culture.

THE ANDROMEDA.

The plants of New England which are most nearly allied to the heath are the different species of Andromeda. These plants vary in height from one foot to seven or eight feet. They resemble the whortleberry in their general appearance, and in their leaves and flowers, but their fruit is a dry capsule, not a berry, and their foliage is not tinted in the autumn. They are, I believe, without an English name. Several species are indigenous in New England, but only two or three of them are common. One of the most beautiful, though extremely rare, is the Water Andromeda, which is found near the edges of ponds. This is the species which suggested to Linnæus the name given by him to the genus. He describes it in his "Tour of Lapland" as "decorating the marshy grounds in a most agreeable manner. The flowers are quite blood-red before they expand; but when full-grown the corolla is of a flesh-color. Scarcely any painter's art can so happily imitate the beauty of a fine female complexion; still less could any artificial color upon the face itself bear a comparison with this lovely blossom." He thought of Andromeda as described by the poets, and traced a fancied resemblance between the virgin and the plant, to which it seemed to him her name might be appropriately given.

One of the most common of our small water shrubs, very homely when viewed from a distance, but neat and elegant under close inspection, is the Dwarf Andromeda. It covers in some parts of the country wide tracts of swampy land, after the manner of the heath, and is not very unlike it in botanical characters, with its slender branches and myrtle-like foliage. It opens its flowers very early in spring, arranged in a long row, like those of the great Solomon's-seal, extending almost from the roots to the extremities of the branches. The flowers all lean one way, each flower proceeding from the axil of a small leaf. Though an evergreen, the verdure of its foliage is so dull and rusty that it is hardly distinguished in the meadows which are occupied by it.

Another remarkable species is the panicled Andromeda, a tall and very common shrub in Eastern Massachusetts, distinguished from the whortleberry by its large compound clusters of densely crowded white flowers of a nearly globular shape. These flowers are much neater and more beautiful on examination than those of the blueberry, and resemble clusters of white beads. They are succeeded by a dry capsular fruit, bearing a superficial resemblance to white peppercorns. The fishermen of our coast have always employed the branches of this shrub, with those of the clethra, on account of their firmness and durability, as coverings to the "flakes" which are used for the spreading and drying of codfish. These two shrubs were formerly distinguished by them as the "black and the white pepper-bush," one having berries of a lighter color than the other.

TREES FOR SHADE AND SALUBRITY.

THE advantages of trees for shade and protection may seem less hypothetical than those we claim for them as agents in nature's economy. Every man clearly perceives that a mere belt of trees will protect his grounds from the severe action of the winds, and shade them from the scorching heat of the sun. This is a point that requires no effort of reason to be understood; it is plain to the senses of every man who knows enough to walk on the shady side of the road for comfort on a summer's day. Even the flocks and herds have mind enough to perceive that a tree will afford them shade, and that the leeward side of a wood will protect them from the wind. But the philosophy even of this branch of dendrology is not fully understood. The extent of the advantages of trees as a protection of the whole country from the force of winds, and their effects upon agriculture and the amelioration of the climate, according to the disposition that is made of them, are hardly appreciated.

It is a foolish canon of taste that substitutes harmony in the disposition of objects in a landscape in the place of that accidental formality in the rows of trees which have grown up spontaneously by the fences in the old farms of New England. This is said to be done by "improvers," to avoid the stiff and checker-board appearance of square fields belted with trees. It is true, that, if we were to look down from an eminence, we should feel more of the sensation of beauty from the view of a landscape in which no such formalities are apparent. These rows make the boundary lines of the different farms and estates and their subdivisions painfully conspicuous. But the disagreeable impressions caused by them are relieved by our sense of the utility and advantages of trees disposed in this formal manner; for by no other arrangement would they afford the adjoining fields equal protection from the winds. If our rustic ancestors had planted all these formal rows of trees and shrubbery which nature has raised in spite of them, they would have proved their wisdom and foresight.

It is often said that solid fences are a better protection from the winds than trees and shrubs. It is true that fences protect those objects that stand on their leeward side, but they aggravate the force of the wind on their windward side. When the wind strikes a solid fence, it creates a forcible eddy; this would be broken and diminished by the action of shrubbery which has no reverberating power. The fence reverberates the wind, the shrubbery absorbs it. If you throw water against a fence, it rebounds with nearly its original force; if you throw water against a mass of shrubbery, there is no appreciable rebound; it enters and penetrates the whole mass. The action of clumps and large groups of trees, especially if they contain their undergrowth, is very advantageous in breaking the force of winds; but of equal quantities of wood and shrubbery, one part disposed in scattered groups, the other drawn out into lines, and standing in the borders of fields, though the first would make a more harmonious landscape scene, the last would be more serviceable to the local climate. Wood in the borders protects the grounds on every side; and so long as the land is divided among the people into small farms, and these farms are also subdivided into fields, we look upon it as expressing the thrift and prosperity of the inhabitants.

Closely connected with the advantages of trees for

shade and protection are those which relate to salubrity; though we must bear in mind that trees are not in all places and situations promotive of health and comfort. It is well known that the inhabitants of our Southern cities, during the sickly season, or from June to October, resort to the "pine barrens" for health and recreation. The air in these woods is perfectly salubrious; even the half-inundated lands communicate no disease when covered with wood. There is something connected with large assemblages of trees that prevents the formation of malaria in the atmosphere, or at least deprives it of its power of communicating disease. Mr. Marsh records the fact, that, in certain unhealthy districts of Italy, "the interposition of a screen of trees preserved everything beyond it, while the unprotected grounds were subject to fevers"; and he adds, "the belief that rows of trees afford an important protection against malarious influences is very general among Italians best qualified by intelligence and professional experience to judge upon the subject."

How many of the seemingly capricious movements of epidemics might be traced to the presence or absence of trees and woods, if our experience were enlightened by special investigation, cannot be determined. An increase of knowledge on this subject may refer these things to some yet unsuspected action of trees in the economy of nature. It is not yet known whether the salubrity of forests be caused by some chemical action of the foliage upon the atmosphere, destroying or absorbing noxious effluvia, or whether the trees act simply as a bulwark against the access of malarious currents of air. Indeed, the principal cause of their salubrity may be the dense mass of foliage with which the ground is covered, especially under a pine wood, preventing the escape of malaria from the soil. An improved knowledge of the chemistry of the atmosphere may at some future time reveal the laws of these influences; and as the chloride of lime disinfects the foul air of a chamber, it may appear that there are emanations from the leaves of trees, combined with their absorbent action, that purify the atmosphere in more ways than have yet been imagined. These hints are only conjectures; but the fact is admitted by the pioneers of the Western States, that the man who cultivates a small clearing in the forest is secure, while the inhabitant of the prairie or of a clearing that occupies the space of several miles is subject to fever and ague and other malarious fevers.

The greater warmth of a wood in summer after dewfall and during all the early part of the night, when people are generally out of doors, contributes very greatly to the salubrity of a forest. The malaria are most pernicious to those who are exposed to the damp chill and dews; and it is well known that those who keep within doors after dewfall in summer are generally exempt from fevers. The temperature of the atmosphere within a wood being warmer than the outer air, whenever the dews are falling, as in the early part of all still summer nights in clear weather, the dwellers in the wood are protected from chills to which the inhabitants of the open districts are exposed in the evening of almost every day. Probably it is a combination of all these circumstances that causes the general salubrity of a residence in a wood; and an increased knowledge of the facts that bear upon this subject would probably multiply our motives for the preservation of the forests.

All these remarks apply chiefly to assemblages of wood of considerable extent. When a few trees surround a house standing on an open plain, they do not produce all the effects we experience in a wood. They are not sufficient to enclose within their area an atmosphere that is not immediately affected by the temperature and moisture of the outer space. The air underneath a small group of trees is not therefore sensibly warmer in the night than outside of them. They are, indeed, for the most part, rather unfavorable to health when so near a house as to generate dampness and prevent the drying action of the sun.

THE ROSE.

In my description of flowering trees and shrubs, I must not omit the Rose, the most celebrated and the most beautiful of flowers; the delight of mankind in all ages and in every country; the pride of all gardens, and the chief ornament of the field and woodside; the poetic emblem of love and the symbol of truth, inasmuch as its beauty is accompanied by the virtues of sweetness and purity. In every language have its praises been sung, and poets have bestowed upon it all the epithets that could be applied to a direct gift from Heaven. From its graces, too, they borrow those images they would bestow upon the living objects of their idolatry. The modest blush of innocence is but the tint of the Rose; its hues are the flush of morning and the "purple light of love." The nightingale is supposed to have become the chief of singing birds by warbling the praises of the Rose, inspired by the beauty of this flower with that divine ecstasy which characterizes his lay. In all ages the Rose has had part in the principal festivities of the people, the offering of love and the token of favor; the crown of the bride at bridal feasts, and the emblem of all virtue and all delight.

So important a shrub as the Rose cannot be an inconspicuous feature either in our wild or our domestic scenery. Every wood contains one or two species in their wild state, and every enclosure in our villages some beautiful foreign roses, which are equally familiar to our sight. I have nothing to say of the multitude of improved varieties lately introduced by florists. There is a point of perfection that cannot be surpassed in the improvement of any species of plant. An additional number of petals does not always increase the beauty of a flower. In the scale of all kinds of perfection, both physical and moral, there is a degree beyond which improvement is only the addition of insipidity.

THE EGLANTINE, OR SWEETBRIER.

The Eglantine is the poetical name of one of the most charming species of rose, generally known in this country as the Sweetbrier, noted for its scented foliage and its multitude of thorns. This species seems to occupy a mean between the tree-roses and the climbers. It often mounts to a considerable height, supporting its position by its thorns. I have seen a Sweetbrier growing wild upon a juniper to the height of fifteen feet, and covering the whole tree. The flowers are small and of a pale crimson, having less sweetness than the common rose. The American Sweetbrier has paler flowers and a smaller leaf; the English plant has larger flowers of a deeper color, and more luxuriant foliage. The American species, however, attains the greater height; it is more fragrant, and more abundant in flowers.

THE SWAMP ROSE.

There is not a sweeter or more beautiful plant, in its native fields, than the common Wild Rose of our meadows. It flowers early in June, clustering in all wild pastures and in all neglected fields, forming beautiful spontaneous hedge-rows by the sides of fences, and groups and beds of shrubbery in all wild lands. The Swamp Rose varies in height, according to the quality of the soil it occupies. I have seen it from four to five feet in height on the alluvial borders of streams, while in uplands it seldom exceeds two feet. This shrub has a fine glossy pinnate foliage, and flowers of a deep crimson, somewhat larger than those of the sweetbrier. Occasionally a variety is seen with white flowers. The Wild Rose is very common near footpaths through the fields, forming natural clumps, often extending into the enclosures of some rustic cottage. In winter it is easily recognized by the fine purple hue of its smaller branches.

But this shrub finds no favor except from the lovers of nature. I have seen men employed in "grubbing up" the Wild Rose bushes that skirted the lanes extending from their enclosures to an adjoining wood. A similar vandalism causes them to whitewash their stone-walls and the trunks of shade-trees, as if beauty consisted in a gloss of art spread over all the works of nature. If we were to carry out the idea of these improvers, we should destroy every wilding in the borders of our fields, and plant florists' flowers in spots of spaded earth cut out of the turf. It is fashion alone that causes the florists' roses to be admired more than the wild roses of the fields and brooksides. They are, it is true, more splendid and full. But who would be pleased to find these petted favorites of gardeners in the rustic lane or the solitary wood-path? Let them continue to be admired in the parterre; but let not our admiration of their artificial beauty cause us to neglect or despise the simple denizens of the field and forest.

WOOD-PATHS.

THERE is no person who is not sensitive to the beauty of a natural wood. All men feel the comfort of its shade and protection, the freshness of its perfumed gales, the quiet of its seclusion, and its many pleasant accompaniments of birds, fruits, and flowers. We do not learn by tuition to appreciate these objects; they are adapted not only to our native wants, but they are the real cause of many of the poetic thoughts and images that abound in all literature. We feel, while rambling under these lofty trees, and over this carpet of leaves and mosses, that nothing which art has accomplished will compare with the primitive works of nature. There is no architecture so sublime as that of a forest; there are no gardens like the little paradises to be found here, wherever accident has left a dell or a dingle open to the sun; there is no music like that of its solitary birds; no worship so sincere as in these temples; no cloistered solitude so sweet as under these shadowy boughs.

Yet how much greater are the charms of a natural wood if it be intersected by wood-paths! When a farmer makes a passage for his wagon through a forest, he operates without artistic design, and his work harmonizes with nature. He thinks only of facilitating progress through his territory; for though he may be alive to all pleasant rural sights and sounds, he cannot pause from his labors to do anything for mere embellishment. He is governed only by his ideas of utility and convenience. Yet the works of decorative art are tame and prosaic by the side of this rude pathway, which has expelled no wild plant from its habitats, nor a single forest warbler from its retreats. We experience within it a true sensation of nature, with a pleasant reminder of simple rural life. It is hallowed by its humble purpose of utility, by its freedom from artifice, by its perfect submission to the care of nature and chance, by its beauty without adornment.

The wood-path becomes henceforth an avenue to all the delights of the season. It introduces us to the productions of the forest in their most interesting condition. The trees that spread their branches overhead shelter it from cold and heat, and permit thousands of beautiful shrubs to grow there that would be fatally crowded in the dense parts of the wood. Multitudes of flowers appear continually in its borders, one host following another in glowing succession, and looking upon us in our journey as with the eyes of so many little sentinels of light and beauty placed here to make the scene delightful to the sense and the imagination. Like birds that multiply around a human dwelling in the wilderness, flowers always become numerous in these woodland paths, and consecrate them to nature.

There is nothing here to call up any disagreeable ideas of pride and pretence, or to excite envy by the ostentatious parade of wealth. Nature never insults the most humble person who enters her sacred precincts. The rich and the poor, the learned and the unlearned, if they have any love of truth and beauty, are equally pleased and instructed. They surrender their hearts to the simplicity of the scenes around them, forget the cares that perplex their minds, and find pleasure in every object they meet. Here are both freedom and seclusion; for though every foot of land has an owner, no invidious signs of appropriation are made apparent to the pilgrim of these walks. Everything has grown up without culture; for these wildings are the flowers that Nature strewed at her feet when she first stepped out of paradise to bless and beautify the earth. No spaded soil about the roots of the flowering shrubs indicates their petted value to some proprietor; no nicely cut turf at the borders of the path shows the exercise of the topiary art, and the consequent exclusion of nature and freedom.

The flowers that peep out from this grassy path and its tangled borders are eclipsed in splendor by the prouder ones of the garden. They are lovely in their wildness and spontaneous grouping; but, like the stars of heaven, they affect the imagination more than the sight. Though fashion may contemn their beauty, nature cherishes and preserves them; and to a poetic eye they have charms that cannot be heightened by art. For everything that blossoms here, or greens the turf, or jewels the trees and shrubbery with purple and scarlet fruit, or scatters incense in their path, was present at the bridal of the earth and sky. The gales that have always swept through these trees are familiar with their perfume; morning and evening greet them, and are acquainted with their beauty; the little brooks know them; sunshine and shadows have played and fondled with them; the wild bee has sipped of their honey, and the birds have nestled in their foliage.

In these fern-embroidered aisles and under these foliated arches, where the birds have warbled ever since the morning stars sung together, here will we linger when we would worship in Nature's sanctuary, and draw from her an inspiration that will make the scenes of earth as delightful as those of romance. We will seek the wood-haunts of the Naiad, where she sits by her fountain, distributing her favors to herb, tree, and flower, and among these dripping dells we will greet her as the "mother of dews." We will drink of her waters with the thrush and the wood-pigeon, and bear home baptismal drops from her well in the leaf-cups of the sarracenia, and incense from her altar in branches of eglantine and sweet-fern. We will sit under these wide-spreading oaks and take our repast with the squirrel, while from the tall treė-top he watches our motions.

We pass, as it were, in a happy dream, through vistas, under tall trees, forming with their foliage and the sky a netted canopy of green and blue, where delicate aerial voices of mingled chirping and song inspire every wanderer with their own cheerfulness. Sometimes there is a stillness almost sublime; in a moment are awakened certain musical and mysterious sounds that fill the mind with dim conceptions of something more beautiful still unseen and unknown; then a confusion of voices without discord; a universal hum, so soft and so melodious that every bird that sings may be distinctly heard above it, his voice made sweeter by this harmonious din. As we view the surface of some still water, embossed with the reflection of embowering shrubbery and of the herbs that fringe the border, the fountain seems to look upon us with distinct vision and to know us. Suddenly we are under the open sky; we have been led out of the wood into the retreat of the hare, who is startled from her repose by our unexpected intrusion.

O happy path to blisses unknown in the outer world! Guide to joys that revellers cannot feel nor the ambitious know. Wherever there is gladness or beauty, or melody of birds and fountains, or little dells full of roses and honeysuckles, or dripping rocks green with velvet mosses and variegated lichens, — to all this wood-path leads the way; now safe through copses of tangled green-brier and clematis; through borders of roses, untrained by art and not planted by man; through beds of raspberries intermingled with ferns, and thickets of tremulous aspens interwoven with sunshine; then under solemn pines, opening into a grander solitude, where dwells perpetual twilight, — halls familiar with darkness at noonday, and visited only by the rays of the morning and evening sun.

Everywhere there is a store of essences on the dewy air; sometimes a scent of pines, such as a mild south-wind at twilight will waft into our windows from a neighboring grove; then the perfume of oaks, less sweet and aromatic, but like that which we may suppose to have surrounded the oracle of Dodona. Now a mild breeze will waft us the scent of strawberry-beds, bearing a message to the bee that tells where the flowers have spread their feast of nectar. At every season the air about these paths is full of sweet odors, that would communicate to our senses the proximity of certain plants. Not a flower appears that does not give some balmy notice of its presence; not a zephyr wanders through this avenue but with wings laden as if it had passed over the plains of Araby.

While strolling in one of these paths, where the ruts of the wagoner's wheels are hardly perceptible along the green turf, I am affected with a glow of pleasure that cannot be felt in a nicely gravelled walk through the grounds of a palace. I feel a sense of tranquil and poetic seclusion here, that would dissolve, as by a spell, at the least appearance of ornamental design. It is difficult to explain the philosophy of this sentiment. But Nature, whose works perfectly harmonize with the rude wood-path and the artless operations of rustic toil, refuses her blessing to the nicely trimmed avenue and the ambitious designs of wealth and pride. In a gravelled walk through a lordly estate there is neither seclusion nor repose; in the pathless wood, seclusion soon becomes painful solitude; but in the unadorned wood-path is sweet retirement, while an endless maze of verdure and flowers renders the solitude charming.

Though the wood-path does not glow with the splendor

and prodigality of a parterre, there is a never-ending variety of objects to enliven the senses and the imagination. Here are sweet violets dotting the greensward with heaven's own azure; roses that breathe into the atmosphere the very aroma of purity; vines that throw their drapery over branches that form our canopy, making the air ambrosial with their fragrant blossoms in summer, and tempting our sight with their purple clusters in autumn. Here are mossy couches so soft, so beautiful, so hallowed, that the young maiden who should sit upon them becomes a goddess; and the student of nature turned pilgrim here would worship her with more devotion than he yields to science.

Take her, thou young enthusiast, and make her the dryad of this wood. Lead her up this rustic avenue, where violets will breathe out their grateful odors to the pressure of her maiden feet. Seat her in the shade of a druidical oak, and fill her lap with roses, which are the symbols of love, and with the flowers of the blue myosotis, sacred to remembrance. Bind her forehead with arbutus, as unfading as amaranth, and bring for her repast strawberries that cluster about these daisied grounds. Then will you feel that mankind are unhappy only as they wander from the simplicity of nature; and that we may regain our lost paradise as soon as we have learned to love nature more than art, and the heaven of such a place as this more than the world of cities and palaces.

THE MAPLE.

IN New England and the adjoining States, the maples are among the most conspicuous and important families of our indigenous trees. Their wood is used for various purposes in the arts, and their product of sugar is of incalculable value. Two of the European maples are cultivated here, distinguished from the American species by their larger leaves and flowers and their darker verdure. I prefer the latter, because they have a smaller leaf, and consequently a more lively and airy appearance, and because they are more beautiful in autumn.

Besides the three most remarkable species in our native woods, there are several smaller maples in New England, not rising much above the height of shrubs, but distinguished by their elegance and beauty. One of the most common of these is the Striped Maple, sometimes called Moosewood. It is a tree of singular grace and beauty, and in Maine and New Hampshire it is abundant, intermixed with the undergrowth of the forest. It is one of the earliest trees in putting forth its flowers. The leaves are large, broad, not deeply cleft, and finely variegated in their tints in autumn. The protection of the forest seems needful to this tree, for it is seldom found among the border shrubbery of fields and waysides. Mr. Emerson thinks it deserving of cultivation. "I have found it," he remarks, "growing naturally twenty-five feet high, and nineteen or twenty inches in circumference; and Mr. Brown, of Richmond, tells me he has known it to attain the height of twenty-five feet. It well deserves careful cultivation. The striking, striated appearance of the trunk at all times, the delicate rose-color of the buds and leaves on opening, and the beauty of the ample foliage afterwards, the graceful pendulous racemes of flowers, succeeded by large showy keys not unlike a cluster of insects, will sufficiently recommend it. In France, Michaux says it has been increased to four times its natural size by grafting on the sycamore."

The Mountain Maple is another small and elegant species of similar habits to those of the Moosewood, being almost entirely confined to the forest, variegated with red and purple tints in autumn. If it is ever seen by the roadside, it is only when the road is bordered by the forest.

THE SUGAR MAPLE.

The Rock Maple is distinguished from the red maple by its larger leaves, which are entire at the margin, and not serrate, having generally three lobes, sometimes five, separated by a smooth sinus instead of a notch. The flowers are greenish, and come out at the same time with the foliage. This tree is larger than any of the other species, it has a more vigorous growth, and affords a denser shade, but it is difficult to distinguish them when divested of their leaves. It is the most abundant species in all the North-eastern States, including the British Provinces, where it serves more than any other tree, except the white pine, to give character to the wood-scenery. It is rare in Eastern Massachusetts, and is not found below this latitude, except among the Alleghanies.

Dr. Rush, speaking of this tree, remarks: "These trees are generally found mixed with the beech, hemlock, ash, linden, aspen, butternut, and wild-cherry trees. They sometimes appear in groves, covering five or six acres in a body; but they are more commonly interspersed with some or all of the forest trees above mentioned. From thirty to fifty trees are generally found upon an acre of land." Major Strickland says of it: "The Sugar Maple is probably the most common tree among the hard-wood species of Canada West. It is found generally in groves of from five to twenty acres; these are called by the settlers sugar-bushes, and few farms are without them."

Though I consider the red maple a more beautiful tree, having more variety in its ramification, and a greater range of hues in its autumnal dress, than the Rock Maple, it must be confessed that the latter surpasses it in some important qualities. The Rock Maple has a deeper green foliage in summer, and is generally more brilliant in its autumnal tints, which, on account of the tenacity of its foliage, last from a week to ten days after the red maple has dropped all its leaves.

THE RIVER MAPLE.

By far the most graceful tree of this genus is the River Maple, to which the cockneyish epithet of "silver" is applied, from the whitish under surface of its leaves. It is not found in the woods near Boston, but is a favorite shade-tree in all parts of New England. It abounds in the Connecticut Valley and on the banks of some of the rivers in Maine. It is rather slender in its habit, with very long branches, that droop considerably in old and full-grown trees. The foliage of this tree is dull and whitish, but it hangs so loosely as to add grace to the flowing negligence of its long slender branches. The leaves are very deeply cleft, like those of the scarlet oak, so that at a considerable distance they resemble fringe; but they are seldom very highly tinted in autumn.

THE DARK PLAINS

CONTAINING MY FIRST IMPRESSIONS OF A FOREST.

In our early days, when all the scenes about us are full of mysteries, and even the adjoining country is an unexplored region, we feel the liveliest impressions from nature and our own imagination. Those who pass their childhood in the woods, and become acquainted with their inconveniences and their dangers, learn to regard them as something to be avoided. The Western pioneer destroys immense tracts of forest to make room for agriculture and space for his buildings. The inhabitant of the town, on the contrary, sees the woods only on occasional visits, for pleasure or recreation, and acquires a romantic affection for them and their scenes, unfelt by the son of the pioneer or the forester. The earliest period of my life was passed in a village some miles distant from an extensive wood, which was associated in my mind with many interesting objects, from the infrequency of my visits. It was at a very early age, and when I first began to feel some interest in natural objects beyond my own home, that I heard my mother describe the "Dark Plains," a spacious tract of sandy country, covered with a primitive growth of pines and hemlocks, such as are now seen only in the solitudes of Canada and the northern part of Maine.

The very name of this wooded region is highly significant and poetical, and far removed from the disagreeable character of names vulgarly given to remarkable places. What eccentric person, among the unpoetic society of

Puritans and pedlers, could have felt sufficient reverence for Nature to apply to one of her scenes a name that should not either degrade it or make it ridiculous! The very sound of this name sanctifies the place to our imagination; and it is one of the very few applied to natural objects, if the original Indian appellation has been lost, that is not either vulgar or silly. Nothing can be more solemn or suggestive, nothing more poetical or impressive, than the name of this remarkable forest.

I attached a singular mystery to this region of Dark Plains. When I first heard the words spoken, they brought to mind all that I have since found so delightful in the green solitudes of nature, — their twilight at noonday; their dark sombre boughs and foliage, full of sweet sounds from unknown birds, whose voices are never heard in the garden and orchard; the indistinct moaning of winds among their lofty branches, like a storm brewing in the distant horizon, sublime from its seeming distance and indistinctness, though not loud enough to disturb the melody of thrushes and sylvias. All these things had been described to me by her to whom I looked, in that early time of life, for all knowledge and the solution of all mysteries. I had never visited a wood of great extent, and the Dark Plains presented to my imagination a thousand indefinable ideas of beauty and grandeur.

It has often been said that the style of the interior arches of a Gothic cathedral was indicated by the interlacing and overarching boughs of the trees as they meet over our heads in a path through the woods. I think also that the solemnity of its dark halls and recesses, caused by the multiplicity of arches and the pillars that support them, closely resembles that of the interior of a forest; and that the genius of the original architect must have been inspired by the contemplation of those grand woods that pervaded the greater part of Europe in the

Middle Ages. The solemn services of the Roman Catholic religion found a people whose imagination having been stimulated by their druidical rites looked upon these wonderful temples as transcending nature in grandeur; and they bowed before the Cross with still greater devotion than they had felt when they made sacrifices under the oak.

There is an indefinable charm in a deep wood, even before we have learned enough to people it with nymphs and dryads and other mythical beings. Groups of trees that invite us to their shade and shelter, in our childhood, on a sultry summer noon, yield us a foretaste of their sensible comfort; and a fragment of wild wood, if we see nothing more spacious, with its cawing crows, its screaming jays, and its few wild quadrupeds, gives us some conception of the immensity of a pathless forest that never yet resounded with the woodman's axe. I was already familiar with these vestiges of nature's greatness, enough to inspire me with feelings that do not become very definite until the mind is matured.

The time had come at last when I was to visit one of these solemn temples of the gods. I was between eight and nine years of age, and was to accompany my parents on a journey from Beverly to Concord, my mother's native town, in New Hampshire. I give this narrative of personal experience, to prove that our love of nature is an innate feeling, which is exalted, but not created, by the imagination. Nothing ever occupied my mind so intensely as the thought of visiting these Dark Plains. Other objects seen on our journey were amusing and attractive; but this wood was the only one that excited in me a passionate interest. All my thoughts were obscure and indefinite, associated with some dreary conceptions of beauty and grandeur; for in our early years we aspire after more exalted feelings than the common scenes of Nature can awaken.

When at length we entered upon the road that led through this forest, the sweetest music had never held me so completely entranced as when I looked up to these lofty trees, extending their branches beyond my ken, with foliage too dense for the sun to penetrate, and all the mysterious accompaniments of the wood, its silence and darkness, its moanings and its echoes. I watched the scenes as we rode slowly by them, — the immense pillars that rose out of a level plain, strewed with brown foliage, and interspersed with a few bushes and straggling vines; the dark summits of the white pines that rose above the round heads of the other species which were the prevailing timber; the twilight that pervaded these woods even at high noon; and I thought of their seemingly boundless extent, of their mysterious solitude, and their unspeakable beauty. Certain religious enthusiasts speak of a precise moment when they feel a certain change that places them in communication with Heaven. If one is ever in a similar manner baptized with the love of nature, it was at this moment I felt that hidden influence which, like the first emotion of love, binds the heart with an unceasing devotion.

I did not at this early age examine individual objects. Yet now and then the note of some solitary bird, or the motions of a squirrel on the outer trees of the wood, held my attention while I was absorbed in a revery of delight. An occasional clearing, containing a cottage with its rustic appendages, opened the sunshine into our path, and made the wood cheerful by this pleasant contrast. When at length we emerged from this gloomy region into the brightness and cheerfulness of the open country, I still dwelt upon the quiet grandeur of its solitudes, and have never forgotten the impressions I had received from them, nor the passionate interest awakened in me before my journey.

About thirty years afterwards I revisited this wood, and traversed the greater part of it, accompanied by an old friend of the generation that had passed before me. From him I learned that the original growth of timber had been mostly felled, and a second growth of inferior height and dimensions occupied its place. He pointed out to me how the whole character of the wood was changed by the simple act of felling the primitive trees. The ground was not so wet as formerly; the standing waters did not occupy so wide a space; the forest contained more openings, the barren elevations not having been supplied with a new growth of trees. In the place of them were a few scrub oaks, some whortle-berry-bushes, and other native shrubs; the trees were smaller, and there was a greater predominance of pitch-pine in all the more sandy parts of the tract, and numerous white birches had sprung up among them.

"Such is the change," he remarked, "which is gradually taking place over the whole continent." He seemed to regret this change, and thought the progress of the civilized arts, though it rendered necessary the clearing of the greater part of the wooded country, ought not to be attended with such universal devastation. Some spacious wood ought to remain, in every region, in which the wild animals would be protected, and where we might view the grounds as they appeared when the wild Indian was lord of this continent. Even at that time I found some acres of forest which had been unmolested still retaining those grand, wild, and rugged features that entitled the region to the poetic name of Dark Plains.

THE RED MAPLE.

NOT dainty of its soil, but thriving equally well in a bog or upon a fertile river-bank, by the side of a stream or upon a dry eminence; coming forth in the spring, like morning in the east, arrayed in crimson and purple; bearing itself not proudly, but gracefully, in modest green, among the more stately trees in summer; and, ere it bids adieu to the season, stepping forth in robes of gold, vermilion, crimson, and variegated scarlet, stands the queen of the American forest, the pride of all eyes and the delight of every picturesque observer of nature, — the Red Maple. There are but few trees that surpass it in general beauty of form and proportion, and in the variety and splendor of its autumnal tints it is not equalled by any known tree. Without this species, the American forest would hardly be distinguished from that of Europe by any superiority of tinting. It stands among the occupants of the forest like Venus among the planets, the brightest in the midst of brightness, and the most beautiful in a constellation of beauty.

The Red Maple is a tree of second magnitude, very comely at all periods of its growth, producing many branches, forming a somewhat pyramidal top while young, but expanding into a round head as it grows old. It is very evenly subdivided, the central shaft seldom being distinguished above the lower junction of its principal branches. The leaves are palmate, of rather a pale green, and the spray, though neat and elegant, does not equal that of the lime or the birch. We associate this tree with the valleys and lowlands, but a wet soil is not necessary for its prosperity. Some of the finest single trees I have known were standing upon a dry soil; but a forest of them is always located in a swamp.

The Red Maple is one of the most common trees in the southern parts of New England, and it occupies a very wide geographical range. In the North it first appears in the latitude of Quebec. It seems to avoid the company of the rock maple, and forms no large assemblages above the northern boundary of Massachusetts, below which the kindred species becomes rare in New England. The Red Maple is abundant in all the Atlantic States, as far as Florida, and there is no other tree that occupies so large a proportion of the wet lands in the Middle States. According to Michaux, it is the last tree which is found in swamps, as we approach the boundary of vegetation.

Preference is generally given to the other two species for planting by waysides and in pleasure grounds in Massachusetts, because they are more luxuriant in their growth. Perhaps they are chosen for the sake of variety, being less common in the woods of this State than the Red Maple; and being planted from nurseries, and costly, they are found chiefly in dressed grounds. But the Red Maple is far more interesting and beautiful than any other species, and its lighter foliage, more airy habit, and more delicate spray bring it into better harmony with wild and rude scenery, as the paler and less luxuriant wild flowers better adorn a wood-path than the more showy denizens of the garden. The Red Maple bears a profusion of crimson flowers in the spring, and from them it derives its name. When the flowers have dropped their petals, the keys, or fruit-pods, that succeed them, retain the same crimson hue for some days, gradually fading into brown as they mature.

SECLUSION AND FREEDOM.

When we go out into the wilds of nature, where the landscape is open and all the works of art are rude and simple, we feel a pleasant exhilaration, not to be explained by referring it to the beauty of the scene. It is a sense of freedom, which is more nearly allied to our love of nature than we are generally aware. The sentiment of seclusion, though a very different mood of the mind, is cherished by similar kinds of natural scenery. If this open landscape be interspersed with fragments of wild wood and sheltered nooks, we perceive its adaptedness both to freedom and solitude. We are not afraid of intrusion, nor do we feel a risk of being intruders. The sight of a few rustic cottages does not clash with these sentiments. But if we suddenly encounter a splendid villa with highly dressed grounds, smooth gravel walks, and other appurtenances of luxurious pride, we perceive our liability of trespassing upon an estate from which, if not excluded, we are sure to be observed. This feeling of restraint destroys our sense both of freedom and seclusion.

It is the power of awaking these sentiments, in connection with others of a poetical cast, that gives their charms to dreary and desolate scenes, to wide wastes of rocky surface containing groups of stunted trees and shrubs. Ocean views are favorable to similar trains of thought, when bounded by an uncultivated shore. But with all the luxury of sentiment inspired by this kind of scenery, no man would consent to be confined to it. Even a long journey through it would be tiresome and depressing. We love solitude and the scenes that inspire the sensation, if our seclusion is to be but temporary. We love to go out and remain a few hours, apart from all appearances of culture, if we know that we may at any hour return from our solitude to the pleasures of home and society. Such was the kind of solitary life that was sought by Thoreau in his hermitage in Walden woods. With all his eccentricities, he was a man of social habits. His love of society equalled his love of solitude, and he sought the gratification of eaeh of these sentiments in his solitary hut, often emerging from it to pass a few hours with his friends in the world which he seemed to have renounced.

The pleasures of forest life are chiefly derived from the sentiments of freedom and seclusion. The woodcutters, though for the most part incapable of feeling the enthusiasm of a poetical mind, find pleasure in the freedom and solitude of a pathless wood, when seasoned with the anticipation of a return to the busy world. Occasionally these foresters are joined by men of culture and sensibility, who in their descriptions of this mode of life dwell with rapture upon the charms of that temporary solitude, and of that freedom which is unrestrained except by the material impediments to their progress. They are delighted with the experience of living apart from the world, mingled with the hope of soon returning to it.

When affected by any absorbing grief, there is something in the rude and desolate aspects of nature that dilutes our sorrows with the romantic sentiments they inspire, while their occasional scenes of beauty afford us a healthful diversion. A flowery and sequestered nook, resounding with the melody of birds and the hum of insects, seems as if designed by some propitious deity for our solace. This seclusion does not clash with our feelings, nor intensify them; it cherishes a train of poetic thoughts, and gives a softer character to our sorrows. I believe we never visit unadorned Nature without gaining some impressions from her scenery that serve to magnify our happiness. In the secluded scenes of the outer world it is not solitude we seek, but a sequestration from all that is wearisome and offensive; and while surrounded by them, our sensation of freedom exalts, as much as that of solitude composes, the mind.

It is the sentiment of freedom that causes the pleasure with which we look upon fields unenclosed and roads not bounded by a fence, but admitting free access to woods and grounds on either side, that seem to invite us to enter. We dislike any manifest signs of the appropriation of earth's green surface. Landscape-gardeners dwell with singular complacency upon the idea of "appropriation," on account of the lordly sense of personal grandeur with which its evidences inspire the owner. They afford a rich man a proud consciousness of his own dignity, by showing him the greatness of his possessions. This is one of the principles of landscape-gardening which is based, not on a sentiment of nature, but of pride, selfishness, and exclusiveness. The expression wrought into landscape by such artifice indicates the stolid feeling of an aristocrat, not the sensibility of a painter or a poet. The effects of working on this principle when improving landscape are offensive to those who would see this earth open to the enjoyment of all rational beings.

We know that all the lands, and the trees and shrubbery that grow upon them, are the property of some legal owner; but we dislike to see the evidence of this paraded before our sight by certain artifices designed for this very purpose. If a number of rustic yeomen and laborers are the owners, who are content to leave them without any ostentatious marks of their ownership, we feel, when rambling in their grounds, the freedom of forest life. In a landscape we would behold a great deal of pasture, of forest, and homely tillage, and but a small proportion of ornamented ground. We perceive the need of fences, and we can easily regard them as objects of beauty, if designed for convenience, and not for show. Dressed grounds not only exclude man, they also banish the wild animals and birds by their nice grading and clearing, and in direct proportion to their extent do they destroy the open expression of freedom in landscape.

The very words employed to designate the different kinds of ground have a poetic or prosaic expression, according as the ideas of freedom and seclusion, or the opposite ones of restraint and exclusiveness, are presented by them. What mind is not agreeably affected by the word "prairie," with its magnificent space and unrestrained journeys, its openness and gladness, its grandeur and its solitude? The words "glen" and "valley," "forest" and "mountain," "field" and "pasture," all awaken images of Pierian freshness and beauty. The word "park," on the contrary, savors less of nature than of the city, less of beauty than of decoration, less of romance and poetry than of taste and artifice; very delightful in a city, with its marble edifices and paved avenues, but in the country like a daub of paint on the cheek of an infant.

THE WHITE BIRCH.

On the sandy plains of many parts of New England, some of the most prominent objects are coppices of slender White Birch trees, intermingled with pitch-pine. These trees are seldom more than four or five inches in diameter, rising to the height of about twenty feet, with a grayish-white trunk, and, as may be observed in winter, a dense and dark-colored spray. This species is called Poplar Birch, from the tremulous habit of the foliage, but is never assembled in large forest groups. Like the alder, it is employed by Nature for the shading of her living pictures, and for producing those gradations which are the charm of spontaneous wood-scenery. In all the Northern States, a pitch-pine wood is generally fringed with White Birches, and outside of them is a still more humble growth of hazels, cornels, and vacciniums, uniting them imperceptibly with the herbage of the plain.

The White Birch is remarkable for its elegance. It seldom divides the main stem, which extends to the summit of the tree, giving out from all parts numerous slender branches, forming a very neat and beautiful spray, of a dark chocolate-color, contrasting finely with the whiteness of the trunk. This tree, when growing as a standard, has more of a pyramidal shape than in a wood; but it does not attain in this country the magnitude of the same species in Europe. The durability of the bark of the White Birch is said to be unsurpassed by that of any other vegetable substance. Selby records a fact related by Du Hamel, which is remarkable. In the ruins of

Dworotrkoi, in Siberia, a piece of birch wood was found changed into stone, while the outer bark, white and shining, remained in its natural state.

So many of the most delightful scenes of nature are in my own mind allied with the different birches, that there is not one that does not immediately call up some charming scenery and impress my mind with pleasant memories. He who in his early days was a rambler in the woods is familiar with the White Birch trees. They have shaded him in his sylvan researches and his solitary musings, his social walks in quest of flowers with the sex for whom the flowers seemed to be created, or with his male companions in pursuit of game. When journeying, these graceful trees, in company with the fragrant pitch-pines, have offered him their flickering shade, and along the sandy plains have defended him from the scorching heat of the sun, and spread a leafy canopy over his rustic path. In the sultry heat of summer noonday, I have often followed the course of some humble cart-path through their tangled undergrowth, gathering wild fruits from bush and bramble, or watching the singing-birds that nestled in their boughs and blended their wild notes with the sound of the green rustling leaves.

All the birches are graceful trees. Their branches are finely divided, like those of the elm and the lime, and many of them incline to a drooping habit. There is a remarkable airiness in their slender feathery spray, rendered still more lively in the White Birch by its small tremulous leaves. This species is found in the highest latitude in which any tree can live. It is the last deciduous tree in the northern boundaries of vegetation in America and Europe, before we reach the Arctic Circle, and the last that appears when we ascend high mountains, occupying the belt just below the line of perpetual snow. It is worthy of notice that the small White Birch in this country, though considered identical with the White Birch of Europe, is greatly inferior to it in size. In America, however, the white canoe birch, a very similar species, equally surpasses the European White Birch. It seems as if the thrifty habit of the canoe birch had some mysterious influence in dwarfing the other species in America.

THE CANOE BIRCH.

Some of the most beautiful assemblages of wood in high latitudes on this continent consist of the Canoe Birch. It is seen in Massachusetts and Connecticut only in occasional groups; but in the States of Maine and New Hampshire, on the sandy river-banks and diluvial plains, it forms woods of great extent and unrivalled beauty. With their tall shafts resembling pillars of polished marble, supporting a canopy of bright green foliage, they form one of the picturesque attractions of a Northern tour. Nature indicates the native habitat of this noble tree by causing its exterior to display the whiteness of snow. The foliage of the Canoe Birch is of a very bright green, and exceeds that of all the family in the depth of its golden tints in autumn. We never see in the foliage of the birches any of that glaucous or pea-green color so common in the maples. The leaves of the Canoe Birch deviate from the ovate form and approach the heart shape. Its bark is almost purely white, and attracts the attention of every visitor of the woods. The clean white shafts of a Canoe Birch wood, towering upward among the other trees of the forest, present a scene with which nothing else is comparable. The uses which have been made of the bark of this tree are so numerous and so familiar to all that it would be idle to enumerate them. Indeed, it would be difficult to estimate its importance to the aboriginal inhabitants of America.

RELATIONS OF TREES TO BIRDS AND INSECTS.

"MY neighbors," said my philosophic friend, "are the cause of more than half the injury my crops receive from caterpillars and other insects. They will not allow the birds a harbor of wood and shrubbery upon their own grounds, and they shoot those which I endeavor to entice by offering them a shelter in my farm. It is strange they cannot understand the mischievous character of their operations of smoothing and grubbing. That little rising ground you see before you, covered with trees and shrubs, is hardly more than a bare rock. It occupies about an eighth of an acre; but no other possible use could be made of it, except as a quarry. The little grove, or coppice, that stands upon it, is the most beautiful object in sight from my house. I have never allowed it to be disturbed or frequented by social assemblages. I keep it sacred for the use of the birds, and it is a perfect aviary. The birds that feed upon the destructive insects that infest my grounds are raised in that temple of the gods, which is watered by numerous little springs that ooze from the crevices of the rock. While they are rearing their young, all species, even if they live exclusively upon seeds after they have left their nest, feed their offspring upon larvæ, which they collect from the nearest ground that affords them a supply. Hence I consider that bare rock, with its trees and shrubbery, the most profitable division of my farm, from the shelter it affords the birds, which are in an important sense my most profitable stock."

I have often thought of my neighbor's remarks, especially when I have observed the diligence of our farmers in destroying upon their grounds every acceptable harbor for the birds. When we are traversing a wood, if we discover an apple-tree growing in a little clearing or open space, we find it invariably exempt from the ravages of the common apple-borer. The same exemption is observed in those fruit-trees that stand very near a wild wood, or any wood containing a spontaneous undergrowth. The explanation of this fact is that the wood affords a harbor to the birds that destroy these insects in all their forms. Orchards and gardens, on the contrary, which are located at any considerable distance from a wood, have not this security. Robins, it is true, are very abundant in orchards, which are their breeding-places; but robins, though the most useful birds that are known to exist, take all their food from the ground. They destroy vast quantities of cutworms and chrysalids buried in the soil, but they take no part of their insect food from the trees. The birds that perform this work are the sylvias, woodpeckers creepers, and other species that live only in woods and thickets. Hence an orchard that is nearly surrounded by a wild wood of much extent is not badly infested by borers and other injurious insects.

All species of insects multiply in cultivated grounds, while the birds, with a few exceptions, that feed upon them, can find a nursery and protection only in the woods. "The locust," says George P. Marsh, "which ravages the East with its voracious armies, is bred in vast open plains, which admit the full heat of the sun to hasten the hatching of the eggs, gather no moisture to destroy them, *and harbor no bird to feed upon their larvæ.* It is only since the felling of the forests of Asia Minor and Cyrene that the locust has become so fearfully destructive in those countries; and the grasshopper, which now threatens to be almost as great a pest to the agriculture of North American soils, breeds in seriously injurious numbers only where a wide extent of surface is bare of woods."

Some men destroy trees and shrubbery in their borders, because they are supposed to harbor insects. But if this be true, it is because they are not sufficient in extent to shelter the birds that feed upon them. The insects that multiply upon our lands deposit their eggs some in the soil, some on the branches of trees and upon fences and buildings. They are nowise dependent on a wild growth of wood and shrubbery. These pests of agriculture need nothing better than the under edge of a clapboard or a shingle whereon to suspend their cocoons or lay their eggs. So minute are the objects that will afford them all the conveniences they need, when hatching and when passing through all their transformations, till they become perfect insects, that no artifice or industry of man can deprive them of their nurseries, or appreciably lessen their numbers. All inventions and appliances used to rid the trees and grounds of these pests never destroyed more than one in a million of their whole number. It is not in the power of man, with all his science, unassisted by birds, to prevent the multiplication of insects from being the cause of his own annihilation. But the farmer, when he destroys the border shrubbery in his fields and the coppice and wood on his hills, exterminates the birds by hosts, while the mischievous boy with his gun destroys only a few individuals. The clipped hedge-row, which is often substituted for a border of wild shrubbery, may assist in breeding insects ; but the birds never build their nests in a hedge-row, unless it be a long-neglected one.

I have in another essay spoken of the scarcity of birds and other animals in the primitive forest. They are not numerous there, because the forest would yield them only a scanty subsistence. The forest border is their nursery and their shelter, but their best feeding-places are the cultivated grounds. There is not a single species whose means of subsistence are not increased by the clearing of the forest and the cultivation of the land; but they require a certain proportion of wild wood for their habitation. Very few species build their nests in the trees and shrubbery of our gardens, unless they are near a wood. In that case the catbird often nestles in the garden, that during the rearing of its young it may be near the grounds that produce larvæ. Most of the woodpeckers, the sylvias, and the small thrushes, including some of our most valuable birds, cannot rear their young except in a wild wood. Yet all these, solitary as they are in their habits, increase under favorable circumstances with the multiplication of insects consequent upon the culture of the soil. It may be affirmed as an indisputable truth, that if their increase were not checked by the sporting habits of men and boys, and the clearing and grubbing habits of "model farmers," birds of every species would increase in the same ratio with the multiplication of their insect food, and proportionally diminish their ravages.

THE BLACK OR CHERRY BIRCH.

THE epithets "black," "white," "red," and "yellow" which are so commonly misapplied to certain trees for specific distinction, — a misapplication very remarkable with reference to the poplar, — are very well applied to the different species of birch, and serve as intelligible marks of identity. The Black Birch, for example, is clothed with a dark-colored bark, which comes nearer a pure black than any other color. No person would dispute the color of the white birches; that of the yellow birch, though not pure, would never be mistaken for anything but yellow; and the bark of the red birch, though nearly white, is so thoroughly stained with red as to demonstrate the propriety of its name.

The Black Birch is also named the Cherry Birch, from the resemblance of the tree to the American black cherry. Its inner bark has the flavor of checkerberry, and its wood some of the colors of mahogany; and it has received names corresponding with these characters, such as Sweet Birch and Mahogany Birch, and was formerly a favorite material for cabinet furniture. The bark of this species and of the yellow birch has very little of that leathery or papyraceous quality which is so remarkable in that of the white birches. This species does not extend so far north as the others, but has a wider geographical range in and below the latitude of New England.

The Black Birch puts forth its flowers very early in the year, of a deep yellow and purple and sensibly fragrant. The foliage also appears early. The leaves are finely serrate, oval, with conspicuous veins, turning yellow in the autumn. Not one of the birches ever shows a tint approaching to red or purple in its foliage. The Black Birch delights in moist grounds, and commonly occupies a stand on mountain slopes and on the banks of rivers. When growing singly on a plain, or in an open space, it takes a hemispherical shape, with its terminal and lower branches drooping to some extent like those of the elm. This tree is conspicuous on craggy precipices, among the mountains, where it extends its roots into the crevices of the rocks, and spreads its branches over chasms and hollows. On these sites it displays a variety of picturesque forms, corresponding with the rudeness and the wildness of the scenery around it. Nature has furnished this tree with a chaffy or winged seed, which is soon wafted and sown by the winds upon mountain-sides and among inaccessible rocks, where the soil collected in thin fissures supplies it with sustenance.

THE YELLOW BIRCH.

The Yellow Birch, named *excelsa* by botanists, from its superior height, is perhaps the most beautiful of the genus. Its branches are extremely numerous, long and slender, corresponding with the superior length of its trunk, and they are prone, like those of the elm, to equality in size, and to divergency from nearly a common centre. Indeed, where this tree has grown as an isolated standard, it commonly displays a very symmetrical head, differing in form from a perfect elm only by less inclination to droop. The leaves of this species have much of the same quality which I have remarked as peculiar to the beech, every leaf standing erect upon its stem. The flexible appearance of the tree is derived entirely from its slender flowing branches.

The Yellow Birch is very abundant in Maine and New Brunswick, and formerly constituted the greater part of the wood which was brought into Massachusetts for fuel. Many of the logs were of immense size before the primitive forest was removed. At the present day we seldom find one more than eighteen inches in diameter, though many slender individuals still occupy our woods. It delights in cold, damp soils, and I have seen the finest standards near springs on an open hillside. The Yellow Birch derives its name from the golden hue of the bark that covers the trunk and larger limbs. This silken bark, which is rolled into multitudes of soft ringlets, is peculiar to this tree.

THE RED BIRCH.

The Red Birch is a rare species, and but very little known. By careless observers it might be mistaken for a white birch, the redness of its bark seeming only a departure from its usual type. The only trees of this species I have seen in Massachusetts were in Andover, in a swamp through which the Shawsheen River flows. If you would behold this tree to the best advantage, you must follow the streams that glide along the level woodlands which are inundated a part of the year. There it may be seen, like some pilgrim bending worshipfully over the stream, by whose beneficent waters it is sustained in beauty and health. Its picturesque attractions, arising from the great variety of its outlines and the peculiar wreathing of its foliage around the stem, are not surpassed by those of the willow, that delights in similar places. The reddish whiteness of the bark and wood has given the name to this tree. It is a tall, bushy tree of rapid growth, rolling up its bark in coarse ringlets, which are whitish with a stain of crimson.

THE INDIAN SUMMER.

When November arrives, leading along with it the short days and the darkness of winter, it opens the windows of the deep woods, pervaded all summer by a sort of artificial twilight. The general denuded state of the forest admits the sunshine into its interior, and brightens it with a cheerfulness exceeding that of any other season. Some light-tinted leaves still remain upon the trees which have been screened by their situation from the frost and the wind, and many an interesting object is exposed to view which was concealed by the foliage in summer. A few asters and gentians still linger in some protected nook, and the chickadees and hemp-birds make the wood lively by their garrulity and their motions. The ground is covered with red, brown, and yellow leaves, making a pleasant carpet for our feet, and increasing all the pleasures of a woodland ramble.

After the fall of the leaf is completed, then, according to tradition, comes the Indian Summer, — a fruitful theme both for poets and philosophical writers, but of which no one knows anything from experience. It may, after all, be only a myth, like the halcyon days of the ancients, the offspring of a tradition that originated with certain customs of the Indian, and which occasional days of fine weather in the autumn have served to perpetuate. It is certain that we have now in the Eastern States no regular coming of this delightful term of mildness and serenity, this smiling interruption of the melancholy days of autumn. We are greeted occasionally by two or three days resembling it after the first cool weather of October, and these short visits are in some years repeated several times. But a true Indian Summer, attended with all the peculiar phenomena described by some of our early writers both in prose and verse, rarely accompanies a modern autumn. It has fled from our land before the progress of civilization; it has departed with the primitive forest. I will, however, for the present, set aside all my conjectures of its mythical character, and treat it as a matter of fact.

The Indian Summer, if such a season was ever known, was a phenomenon produced by some unexplained circumstances attending the universal wooded state of the country that existed for many years after its settlement. According to the most apparently authentic accounts, it did not arrive until November, nor until a series of hard frosts had destroyed all the leaves of the forest. It then appeared regularly every year. At the present time people know so little about it that they cannot name the period of the autumn when, if it were not a thing of the past, it should be expected. Will the disappearance of this phenomenon admit of a philosophic explanation? Let us consider some of its probable causes, and the effects of the changes which have taken place in our land.

It has been observed that a meadow covered with luxuriant grass and other herbage cools the atmosphere that rests upon it much more rapidly than a similar meadow covered with a scanty herbage. The moisture exhaled into the air by vegetable perspiration is greater than from any other natural surface; and as the radiation of heat is rapid in proportion to the moist condition of the atmosphere, the cooling process over a grassy meadow is vastly greater than over a similar ground bare of vegetation. A wood, in like manner, by exhaling through its foliage the moisture it draws from the earth, cools the atmosphere in proportion to the amount of its foliage, while at the same time it shades the ground from the sun. Anything that should check this vegetable perspiration would in the same ratio preserve the heat of the atmosphere by diminishing the radiation of heat that takes place more slowly in dry than in moist air.

This is precisely what happens soon after the first severe frosts of November, when the whole extent of the forest over thousands of miles is laid bare in the brief space of two or three days. There is a sudden and universal diminution of the moisture that was given out from the leaves of trees and other plants before the frost had destroyed them; for the evaporation caused by the drying of fallen leaves and herbage is comparatively slight, and ceases after a few hours' exposure to the sun. The atmosphere being dry, and the radiation of heat proportionally small in quantity, all these circumstances, if no unusual atmospheric disturbances occur from any other hidden cause, unite in producing a sudden and universal accumulation of heat. The warm period that follows is the Indian Summer.

A writer in "Silliman's Journal" of 1833, who advances a very different theory to explain this phenomenon, makes a statement that favors my view: "It appears to us that the existence and duration of the Indian Summer in this country has an important connection with the extensive forests and uncultivated lands peculiar to America. And it is worthy of remark, that, according to the recollection of the oldest of our inhabitants, its former duration was often three or four weeks; whereas its present continuance is short and uncertain, seldom exceeding ten or fifteen days. It appears also that this decline has been somewhat regular, keeping pace with, and evidently influenced by, the gradual uncovering of the country."

It is surprising that the writer, after making these observations, should resort to some unintelligible reasoning about the trade-winds, and certain assumed electric phenomena, to account for the Indian Summer. I can easily believe that before the encroachments upon the American forest were very extensive, this halcyon period of autumn may have occurred every year with great regularity. But since the clearing is almost universal, these conditions have been entirely changed. During the primitive state of the forest, its sudden denudation produced a more complete revolution on the face of the country than could possibly happen at the present time. The clearing of the woods has also cast down the barriers that impeded the circulation of the winds; at present these winds, sweeping freely over the continent, would counteract any influences, whatever they might be, that would produce an Indian Summer in any locality.

The true Indian Summer was a period of very mild weather, lasting from ten to fifteen days, and accompanied neither by wind nor rain. It has been incorrectly described by certain writers as attended with fog. The sky, though somewhat dim, was not obscured by vapor, but by a sort of ruddy haze, that veiled the prospect, as it often will during a series of warm, still days happening at any season. I draw my inferences from what I have reason to consider the most authentic accounts. The air was dry; and it could not have been otherwise. If it were moist, the increased radiation would soon dissipate the heat and put an end to the Indian Summer, which was never known to survive a copious and extensive fall of rain. The atmosphere was described as being obscured by smoke, rather than vapor, and this was most apparent in the latter part of the day. This smoky atmosphere has led some writers to suppose the whole phenomenon to be caused by fires in the woods.

According to tradition, no part of the year was more delightful than this short period. Those accounts, however, that extended its duration beyond the space of fourteen or fifteen days were undoubtedly exaggerated. The nearest approaches to an Indian Summer which I have witnessed in its proper season have never lasted a week. In our day, when a warm week occurs in the autumn, it comes at no regular or expected time. This irregularity of its occurrence proves that it is not to be identified as the Indian Summer, which was regular in its happening immediately after the entire denudation of the forest. Similar but shorter periods of mild and serene weather may happen, at the present epoch, in winter and spring as well as in autumn. These irregularities of the weather cannot be explained; nor can we make predictions of the time when any of them may happen. But a warm period in October or December or January is not an Indian Summer, which belonged to November, and is only a relic of the past.

The origin of the name is explained by Dr. Lyman Foot, in the third volume of "Silliman's Journal." He says: "If you ask an Indian in the fall when he is going to his hunting-ground, he will tell you when the fall summer comes, or when the Great Spirit sends our fall summer; meaning the time in November which we call the Indian Summer. And the Indians actually believe that the Great Spirit sends this mild season in November for their special benefit."

THE POPLAR.

In the latter part of April, some of the most conspicuous groups in many of the wooded districts of Northern New England are Poplar woods, full of olive-green aments, and giving the hue of their blossoms and of their pale green spray to large portions of the forest in scattered assemblages. At this period the poplars are an important ingredient in our wood-scenery, especially as their colors vary considerably from those of other trees until all kinds are in full foliage. They have the merit also of preceding a greater part of the forest in the development of their flowers. The aments of a few species are variegated with red and purple stamens; but the generality do not vary from a pure olive. The Poplar has not many of the qualities of a beautiful or picturesque tree. It is marked by a coarse and straggling spray, without any variety in its combination. It is deficient in beauty and density of foliage, which is chiefly remarkable for its fragrance and tremulous habit.

All the poplars are rapid in their growth, and will prosper in almost all situations. They prefer a moist, sandy soil, but shun the peat meadow. Their rapidity of growth renders them valuable where a speedy plantation is wanted. Hence they are very generally planted by the sides of dusty thoroughfares, not being dainty in their choice of soil and situation. The species generally employed for such purposes is the Abele, or Silver Poplar, which possesses these requisite properties in a higher degree than our native trees. It displays also more beauty of foliage, and takes a rounder and handsomer shape than most others. One of the defects which I have frequently observed in the shape of the large poplars is a leaning of the branches rather awkwardly toward the south-east, caused by the prevalent north-west winds acting upon branches of great proportional length, and possessing very little elasticity. This inclination is observed more or less in other soft-wooded deciduous trees.

THE CANADA POPLAR.

The Canada or Balm of Gilead Poplar is more frequent by our waysides than any other species. It is a tree of the first magnitude, attaining a great size in the bole as well as a superior height. It is distinguished by its large leaves, of a bright glossy verdure, and its long branches, always subordinate to the central shaft, which may be traced nearly to the summit of the tree. Before the leaves begin to expand, the buds are covered with a yellow glutinous balsam, that diffuses a peculiar and very penetrating but agreeable odor, unlike any other. Sir John Franklin remarks that this tree constitutes "the greatest part of the drift timber observed on the shores of the Arctic Sea." It has a very wide geographical range, extending from Canada to the Missouri River, and is in many places called the Ontario Poplar. It is abundant in the northern woods, but is found in the southern parts of New England only by the roadsides and in the enclosures of dwelling-houses. The balsam is gathered in all parts of the country as a healing anodyne, and for many ailments it is a favorite remedy in domestic medicine; but no place has yet been assigned to it in the pharmacopœias. All the poplars produce more or less of this substance. It is very different from turpentine, more agreeable when perceived in the air, but pungent and disagreeable to the taste.

THE BLACK POPLAR.

There are several of the poplars that are not easily distinguished, and the different and various accounts of them by botanists have increased this confusion. Part of the difficulty arises from the diœcious character of the poplar, causing in some instances the male and female trees to be mistaken for different species. This is particularly remarkable in the Balm of Gilead poplar. The female tree is smaller than the male, with larger leaves, and annoys us by the abundance of cottony down that covers the ground for a considerable space around it. The male tree is taller and more spreading, and would hardly be recognized as the same species.

The Black Poplar is often planted by roadsides with the Canada poplar, and may be distinguished from it by the greater elegance of its proportions, its smaller foliage, and, when in flower, by its reddish and purple catkins. It is preferred to other species on account of an inferior tendency to that suckering habit which is so disagreeable in the poplar tribe. It seems to me that no persons who should see the Canada poplar and the Black Poplar growing side by side, would hesitate in giving preference to the latter, which is in almost every point a more beautiful tree.

This species is called in Europe the Athenian Poplar. According to Selby, "the classic appellation of Athenian Poplar led to the supposition in England that this species is indigenous to Greece, and that it derived its name from the city of Minerva. Several learned botanists were misled by this name; but it was finally ascertained that North America is its native country, and from its abundance in a particular township called Athens it received the imposing title of Athenian Poplar."

THE RIVER POPLAR.

The River Poplar is not rare in the New England forest, but it is little known as an ornamental tree. Emerson says: "It is much the tallest and most graceful of those which grow naturally in New England. Its foliage is equal to that of the Balm of Gilead in size, and superior to it in depth of color; and the abundance of its aments in the spring, and the rich colors of its leaf-stalks and young branches, when growing in somewhat dry situations, make it a beautiful object." The aments of this tree are not olive-colored, like those of the two aspens, but inclining to red, though not so bright as those of the black and Lombardy poplars. It is very justly called the River Poplar, being found chiefly in wet places, near brooksides, on the banks of rivers, and in alluvial valleys which are liable to be inundated in spring. This tree displays the characteristic peculiarities of the family in giving out its lateral branches at a sharp angle and subordinate to the trunk.

SOUNDS FROM TREES.

"THE earliest chant," says Momsen, "in the view of the Romans, was that which the trees sang to themselves, in the green solitudes of the forest. The whisperings and pipings of the *favorable spirit* in the grove were repeated by the singer, with the accompaniment of the pipe." Certain trees belonging to the sacred groves gave oracular sounds, which were interpreted by musicians, and received by all men with faith and reverence. From the earliest ages men have listened to sounds from trees as music and as the voice of some deity, affording auguries of future events; for, as they reasoned, if a deity speaks to us, what sounds would be a more appropriate medium of communication than those of the trees which formed their temples and their altars? The sanctity attributed to certain groves by the ancients was probably owing to some peculiar sounds emitted by the trees, no less than to the grandeur and impressiveness of their assemblages.

Every tree, when swept by the winds, gives a sound in harmony with the character of its leaves and spray. The sounds from the lofty branches of firs and pines remind the listener of the murmuring of waters, and inspire the most agreeable sensations. How often have I sat under the shade of a pine wood, and listened to the fancied roaring of the distant waves of the sea, as the winds passed through their foliage. When the breeze commences, we hear the first soft rippling of the waves; as it increases, succeeding waves of fuller swell flow tremulously upon the strand, and as the wind subsides melt into silence as they recede from the shore. Other trees produce very different sounds. The colors of their leaves, and the glittering lights from their more or less refractive surfaces, do not differ more than the modifications of sound drawn from them by the passing winds. Every tree is a delicate musical instrument, that reminds us of the character of the tree and the season of the year, from the mellow soothing tones of willow leaves in summer to the sharp rustling of the dry oak-leaf that tells of the arrival of winter.

The sounds from trees are a very important part of the music of nature; but their agreeableness comes rather from certain emotions they awaken than from the melody of their tones. Nature has accommodated her gifts to our wants and sensibilities, so that her beneficence is never so apparent as in the pleasures we derive from the most common objects. If we are afflicted with grief or wearied with care, we flee to the groves to be soothed by the quiet of their solitudes, and by the sounds from their boughs which are tuned to every healthful mood of the mind. Among the thousand strings that are swept by the winds, there is always a chord in unison with our feelings; and while each strain comes to the ear with its accordant vibration, the mind is healed of its disquietude by sounds that seem like direct messages of peace from the guardian deities of the wood.

We find in the works of Ossian frequent allusions to the sounds from trees, to heighten the effect of his descriptions. As the "Spirit of the Mountain," he addresses the wind that bends the oaks, and gives out that deep melancholy sound that precedes a storm, "when Temora's woods shake with the blast of the inconstant winds." He speaks of the "sons of song" as having gone to rest, while his own voice remains, like the feeble sounds of the forest, when the winds are laid. When the aged oak of Morven bends over the stream, its sounds are mournful, like those of a harp when swept by the wind. According to Ossian, it is the oak that blends its music with the sounds of lamentation, and sings the dirges of departed heroes. And the bard declares that he will cease to mourn for them only when the music of the oak shall no longer be heard in the groves of echoing Cona.

When a strong wind prevails, the leaves of all trees are put in motion, and their sounds cannot be distinguished; and during a storm the roar of winds among their branches is almost deafening. This is the grand chorus of the elements; but the sounds that affect us most agreeably are such as come from light movements of the wind and harmonize with the warbling and chirping of birds. It is the aspen that gives out those lulling melodies that spring from the gentle gales of summer. When we are sitting at an open window on a still evening, or sauntering in a wood, or musing in the shade of a quiet nook, when the wind is so calm that the hum of the invisible insect-swarms, hovering in the air, is plainly audible, then is the trembling motion of the aspen leaves peculiarly significant of the serenity of the elements. They produce a tranquillizing sound, associated with rest in the languor of noonday, or with watching in the still hours of a summer night.

When the quiet of the atmosphere begins to yield to the movements of a rising tempest, the aspen, by its excessive agitation, gives prophetic warning of its approach. Often, in a sultry evening, the first notice I have received of a rising thunder-storm came from the increased trepidation of an aspen that stood before my window. So delicate and sensitive is the foliage of this tree that it is excited to action by atmospheric changes before that of any other tree is moved. Thus, while the rustling of the aspen leaf, when gentle, indicates the tranquillity of summer weather, there is likewise an expression of melancholy in its tones when more severely agitated, that forebodes a general stirring of the winds as they come up from the gathering-place of the storm.

I have spoken only of those sounds from trees which are caused by the action of the winds upon their leaves and branches. But there are incidental sounds belonging to the woods, which are modified so as to produce feelings awakened by no other situation. It is in the deep stillness of the forest, and over spacious and uninhabited plains, that we feel most sensibly the peculiar effect of bells, whether it be the solemn peal of a bell from a church tower or the tinkle of a cow-bell that reminds us of simple rural life. The ordinary toll of bells is much more impressive than a chime in these solitudes, because the artificial melody of the chime does not so agreeably harmonize with natural sounds.

In winter the sounds from trees, except in a pine wood, are greatly modified by the absence of foliage. It is at this season, therefore, that we pay the most attention to incidental sounds. When the snow upon the ground has been hardened by repeated freezing and thawing, I have often chosen this occasion for winter rambling in the woods. The loneliness inspired by their seclusion is never so keenly felt as at this season, when there are but few sounds from birds and insects. Then does the stroke of the woodman's axe affect us with the most cheerful emotions. It reminds us of the presence of other human beings in the wood, and enlivens the solitude, as the sight of a little cottage in a wilderness affords the traveller a sensation of the joys of home.

The reverberations of the forest are most remarkable in lonely places, where the silence is favorable to their distinctness. It is by means of echoes that Nature appropriates all artificial sounds, and makes them a part of her own harmony; and, by a little reflection, we shall discover that echoes are a part of the universal life of nature. It is in winter, however, that they most sensibly affect us. In summer we feel that we are not alone; for millions of voices declare the presence of innumerable happy creatures, chirping and singing around us. In winter these voices are mostly silent. It is then that these invisible deities, who were supposed by the ancients to dwell in hidden places in the form of beautiful nymphs, return cheerful responses to all sounds that are awake. When the solitary woodman strikes his axe at the root of the tree, his benevolent echo responds to the sound, reminding him that he is not alone; and the consciousness of this presence animates him to more cheerful exertion.

THE LOMBARDY POPLAR.

There are not many trees that take the shape of a long spire; but Nature, who presents to our eyes an ever-charming variety of forms as well as colors, has given us this figure in the arbor-vitæ, the juniper, and the Lombardy Poplar. This was the species which was cultivated by the Romans, the classic Poplar of Rome and Athens. To this tree Ovid alludes when he describes the resinous drops from the Poplar as the tears of Phaeton's sisters, who were transformed into poplars. Smith says: "Groves of poplar and willow exhibit this phenomenon, even in England, in hot calm weather, when drops of clear water trickle from their leaves like a slight shower of rain."

The Lombardy Poplar is interesting to thousands in this country, who were familiar with it in their youth as an ornament of roadsides, village lanes, and avenues. It was once a favorite shade-tree, and still retains its privileges in some ancient homesteads. A century ago, great numbers of Lombardy Poplars were planted by village waysides, in front of dwelling-houses, on the borders of public grounds, and particularly in avenues leading to houses that stand at some distance from the high road. A row of these trees is even now suggestive of an approach to some old mansion, that still retains its primitive simplicity.

Great numbers of Lombardy Poplars were destroyed at the beginning of this century, from the notion that they generated a poisonous worm or caterpillar. But some of these ancient rows of poplars are occasionally seen in old fields where almost all traces of the habitation they accompanied are gone. There is a melancholy pleasure in surveying these humble ruins, whose history would illustrate many of the domestic habits of our ancestors. The cellar of the old house is now a part of the pasture land; and its form may be dimly traced by an angular depression of the surface. Sumachs and cornel-bushes have supplanted the exotic shrubbery in the old garden; and the only ancient companions of the Poplar now remaining are a few straggling lilacs, some tufts of houseleek, and perhaps, under the shade of a dilapidated fence, the white Star of Bethlehem is seen meekly glowing in the rude society of the wild flowers.

But the Lombardy Poplar, once a favorite wayside ornament, a sort of idol of the public, and, like many another idol, exalted to honors beyond its merits, fell suddenly into contempt and neglect. After having been admired by every eye, it was spurned and ridiculed, and cut down in many places as a cumberer of the ground. The faults attributed to it were not specific defects of the tree, but were caused by a climate uncongenial to its nature. It was brought from the sunny clime of Italy, where it had flourished by the side of the orange and myrtle, and transplanted to the snowy plains of New England. The tender habit of the tree made it incapable of enduring our winters; and every spring witnessed the decay of many of its small branches. It became prematurely aged, and in its decline carried with it the marks of its infirmities.

With all these imperfections, it was more worthy of the honors it received from our predecessors than of its present neglect. It is one of the fairest of trees in the greenness of its youth, far surpassing any other poplar in its shape and in the density and general beauty of its foliage; but nearly all these old trees are gone, and few of the same species are coming up to supply their places. While I am writing, I see from my window the graceful spire of one solitary tree, towering above the surrounding objects of the landscape. It stands there, the symbol of decayed reputation; in its old age still retaining the primness of its youth, neither drooping under its infirmities nor losing in its decrepitude the fine lustre of its foliage. In its disgrace, it still bears itself proudly, as if conscious that its former honors were deserved, and not forgetting the dignity that becomes one who has fallen without dishonor.

There is no other tree that so pleasantly adorns the sides of narrow lanes and avenues, or so neatly accommodates itself to limited enclosures. Its foliage is dense and of the liveliest verdure, making delicate music to the soft touch of every breeze. Its terebinthine odors scent the vernal gales that enter our open windows with the morning sun. Its branches, always turning upwards and closely gathered together, afford a harbor to the singing-birds, that make them a favorite resort; and its long, tapering spire, that points to heaven, gives an air of cheerfulness and religious tranquillity to village scenery.

THE TROUT-STREAM.

I HAVE never been a zealous or a diligent angler, and whenever I have thus employed myself, it was rather as a voluptuary bent on the quiet observation of nature, than as a lover of the sport. Yet I will confess that next to rambling through a wood-path or over an old by-road in the country, I cannot name a more delightful journey than that of following a trout-stream, especially if engaged in the pleasant occupation of trolling for the timid tenants of the brook. The angler passes down the stream; and seldom in this direction, save when it is lost in a wood or a fen, will it disappoint his pursuit. Its intricacies are a source of constant amusement, and its momentary disappearances serve but to awaken our interest. While moving with the stream, it can never entirely elude our observation. But if we turn the opposite way, and try to discover its source, we soon become involved in the perplexity of the metaphysician when he endeavors to unravel the mystery of final causes.

But there is a peculiar excitement attending a search for the original source of the stream, that has often tempted us to seek for it. We imagine it is some shady nook or dripping dell, in which the ferns and mosses have their paradise; and that, if we could but gain this spot, we should view the sacred urn of the Naiad, and observe how she distils its waters from the dews of heaven and the dappled clouds of morning. We wander through glens and thickets and over plains and valleys, pausing only to note the flowers of every hue

that mark its devious course by their greater profusion, and the birds of various plume that come to dabble in its waters. All the minor beauties of nature attend its course, and seem like an enchanted procession following its sparkling light. Hope awakens with every new accession of beauty, and as the stream ripples over the uneven surface, we are led on by its delusive melody to more earnest pursuit. The waters soon divide, and, while we are lured by the most apparent flow, they divide again, and reveal their course only by an occasional glimmer. Our feet soon become embedded in mosses and aquatic plants, and we see before us steep declivities covered with spleenworts and enamelled with flowers. The verdure and the glow, the beauty and the magnificence of foliage and vinery and flowers and evergreens clustering at our feet and depending from the trees and cliffs, seem to assure us that we are near the birthplace of the stream. But it is not there; the urn of the Naiad is still no nearer than ever, and the pleasant gurgling of waters higher up the declivity is but the song of creation, which has had no beginning and will never cease.

But when we renew our trolling down the stream, in this direction there is sufficient intricacy to excite even the dullest imagination. The stream always seems to take that course which is most difficult to follow, — over bogs and through narrow ravines, along under steep rocks embroidered with ferns on the one side, and boulders and gravel heaps on the other. Then it suddenly widens into a broad sheet of water, so shallow that we can only see its sparkle under the tall grass and rushes, variegated with frequent groups of blue iris and clumps of nodding sarracenia. Some plats of tall rushes show by a narrow and irregular parting where the channel of the stream may be traced; and in order to reach the point where it once more flows over dry land, we remit our task of angling. We pass along the edge of a meadow in a foot-path under a grove of maples, through tangled beds of silver-weed, startling now and then a peetweet and sometimes a solitary heron.

There are but few countries whose surface, in those parts which are not mountainous, is so wild, so irregular, and so favorable to picturesque views as New England. Instead of gently rolling grounds, we find hills of abrupt forms, or, if rounded, rising suddenly out of levels. When these valleys are cleared and drained, they become productive farms; and the little settlements upon them, a mill turned by the waters of a brook, and the surrounding hills covered with a various growth of trees and shrubs, are singularly romantic. We seldom follow a trout-stream half a day's journey without passing one of these mill-seats and its few adjacent farm-houses. The stream is widened into a mill-pond above the dam, and becomes one of the minor attractions of our journey. When viewed with the grove of coppice that surrounds it, it seems a living sapphire in an emerald setting. On meeting with one of these interruptions, you may perceive that the trout-stream has for some distance been straightened into a canal, and serves to drain the land of an adjoining farm. You follow this through a path over the black soil, beside a spontaneous border of roses and azaleas, till you are led again into the forest.

Here the stream runs riot through moss-covered boulders, and round the trunks of trees, over ledges, and through beds of trillium and meadow cowslips, then creeps along the margin of a fen sweet with wild-strawberry plants, and suddenly leaping across a rustic lane, under a rude bridge of planks, it pours down a little cascade into a widening shaded by a few willows. This has one deep side, where it gurgles under the overhanging roots of the trees, and a shallow side with a gravelly shore that serves as a watering-place for cattle and flocks. Here it takes a momentary rest; then leaps forward tumultuously through a glen bordered with alders and honeysuckles, occasionally glittering in sunshine in the openings, like a frolicsome child who often turns beaming with laughter. Then we trace its quiet meanderings through a wide level of green meadow, impurpled with the blossoms of pea-vines, and where Arethusa, once the nymph of a fountain, scatters her bloom over the meadow like wreaths from the rainbow.

But it would be vain to bring to memory all the green lanes we have crossed, in following the capricious stream in its wanderings, of all the sweet and flowery meadows we have passed over, of the dank, rushy shallows we have waded, of the tracts of dark, silent woods through which we have followed it, and of the numerous cascades it has formed as it leaps down from the table-land into the space below. It would seem as if it consciously pursued the most picturesque paths over the country, affording glimpses of distant towns, when suddenly emerging from the hills, then leading us almost to the doorstep of rustic farm-houses, surrounded by their solemn cattle and their smiling children.

The day begins to decline as weariness creeps over us. The outlying fields show but narrow gleams of sunshine between the gathering shadows. The brook still keeps on its restless and melodious course, not ceasing its motions with the sleep of animated nature, nor its music with the silence of the birds. The trees grow dim and dubious in the shade of the hills, while some of their loftier summits are tipped with the amber glow of sunset. Homeward we take our solitary walk, while the vesper-bird sings from some neighboring hay-field, or, still later, the whippoorwill chants his melancholy notes as we wend our way through dewy footpaths to our home in the village.

THE ASPEN.

All lovers of nature admire the Aspen on account of its name, which, like that of the willow, is poetical, both from its musical sound and from association. There is no tree more celebrated in emblematical literature than the Aspen. Its sensitiveness to the least movement of the wind, its restless motions, as if some morbid occasion of disquiet unceasingly attended it, have given it a place in the poetry of all nations. But setting aside its symbolical meanings, its suggestions of fickleness and caprice, of levity and irresolution, of impatience and instability, and the use that has been made of it in satirical writings to symbolize the "inconstant temper of woman," the beauty and motion of its foliage alone would always attract admiration. As the Aspen is the only tree whose leaf trembles when the wind is apparently calm, its gentle rustling is always associated with still summer weather.

THE GREAT AMERICAN ASPEN.

The Great American Aspen is a remarkable tree. In height it is unsurpassed by any of the poplars, though there is little about it that is attractive except its great height and its peculiar foliage. It is seldom of large dimensions, and it is without symmetry or elegance in its ramification. Its branches seem to have a straggling growth, not extending so widely, nor at so acute an angle, as those of the poplar. Its foliage is its principal ornament. This would be very dense if it were not for the scarcity of small branches, which are so far apart as to give the tree a meagre appearance, even when full of leaves. The leaf is beautiful, being round ovate, deeply serrate, and put in motion by the slightest breeze. As a standard the Great Aspen is not highly prized. It is easily broken by the wind, and is without symmetry, — a necessary quality in a tree of the poplar tribe, which possesses none of the properties of grandeur. But when the trees of this species form a dense wood, they are unsurpassed in the beauty of their perfectly straight shafts, with their smooth, greenish bark extending upward to an immense height without a branch. The Great Aspen is very common in the woods of Maine and New Hampshire, where the second growth of timber predominates.

The specific name of this tree, *grandidentata*, always affected me ludicrously, when I considered that it was applied to it merely from the deep indentations on the edge of its serrate leaves. *Excelsa* would be a more appropriate name for the species, on account of its superior height.

THE SMALL AMERICAN ASPEN.

This tree resembles the great aspen in almost every particular except size. It is a very common tree in our woods, but is so little esteemed that it has received no protection and is seldom planted by our roadsides. It is found chiefly in copses on the sides of some gravelly bank, growing almost alone, with a few cherry-trees and white birches, and an undergrowth of brambles and whortleberry-bushes. It is often abundant on little dry elevations that rise above an oak wood standing on a clay level. It is remarkable for its slenderness of habit and the smoothness of its pale-green bark, which becomes whitish and rough as the tree grows old. Its principal defect is the thinness of its foliage and spray; its small branches are few and far apart, and its leaves small and sparse. Yet the beauty of each individual leaf is unrivalled. It is heart-shaped, finely serrate, and when young is fringed with a soft, silky, and purple down. It would be difficult to select a branch from any other tree, when in leaf, so beautiful as a spray of the Small Aspen.

I do not understand the botanical difference between the Aspen and the poplar, except that the former includes certain species that possess in an exaggerated degree the family characteristic of a tremulous leaf. The Aspen, however, is the proverbial tree, the tree of romance and fable. Hence we regard it with more interest, though in America the two aspens fall short of the poplars in almost every point of elegance and beauty.

RELATIONS OF TREES TO POETRY AND FABLE.

FROM the earliest period of history, mankind have looked upon trees and woods with veneration, regarding them as special gifts of the gods to the human race. The ancient priests and philosophers used them as their places of retirement, both for the study of wisdom and the services of religion. Hence arose that early custom of planting trees in circles, forming a kind of amphitheatre, for religious assemblies. The teachers of philosophy used the same circular groves. These were held in the greatest reverence; and no man dared to commit the sacrilegious act of cutting down any part of them or defacing any of the trees. By means of these circular groves, wise and holy men obtained that seclusion and quiet which it was not easy to find in towns and cities. They were both schools and chapels, devoted to religion and philosophy. Hence the often-quoted remark of Pliny that "the groves were the first temples of the gods."

It is not improbable that many of the ancient superstitions relating to trees and groves originated with wise men, who believed that such holy fears alone would restrain the people from devastating the whole earth by the destruction of trees. Science now supplies mankind with rational motives for their preservation, in place of the religious scruples of ancient communities. I am inclined to believe that many a rational principle has been advocated by wise men under the guise of theology. The druidical priesthood foresaw that the oak, from the superior value of its timber, could not be saved from the woodman's axe except by certain ceremonies on their part that should render it sacred in the eyes of the people. To impress this idea of its sanctity upon their minds, they made use of its leaves and branches to consecrate all important private or public transactions.

In still more ancient times, the priests adopted the expedient of dedicating to some one of the gods, particularly to Jupiter, certain woods and groves, which were thenceforth held in veneration by all men, including even invading armies, whose chiefs, while respecting neither the lives nor the property of the enemy, held these consecrated groves sacred and inviolable. Hunting was forbidden within them by this superstition, and its injunctions were in all cases religiously observed. It is even asserted that the wild animals in these sacred groves had become so tame, from the permanent security they enjoyed, that they did not flee from the presence of man.

Many persons formerly believed that trees felt the stroke of the woodman's axe, which disturbed the repose of some resident spirit. The ancient Greeks supposed certain trees to be inhabited by wood-nymphs, and that these deities uttered groans when the axe was laid upon the tree. These sounds gave origin to the sacred oracle of Dodona. There were two kinds of nymphs supposed to inhabit trees, — an inferior class that lived during the life of the tree, and died when it perished; and a superior class, like the dryads, who could pass at will from one tree to another. "One might fill a volume," says Evelyn, "with the history of groves that were violated by wicked men who came to untimely ends; especially those upon which the mistletoe grew, than which nothing was reputed more sacred."

The custom of planting a tree at the birth of a child has prevailed among certain nations from the earliest times, and is still observed in some parts of Europe.

Connected with this custom was the idea that the fate of the child was mysteriously associated with that of the natal tree, which created the strongest motives, arising from parental affection, to preserve the tree, and on the part of the child to protect it when he attained his manhood. Nothing is more evident than the beneficial tendency of all these superstitions, at an early age of the world, when men were not wise enough to be governed by the principles of reason and science.

The ancients placed the Naiad and her fountain in the shady arbor of trees, whose foliage gathers the waters of heaven into her fount and preserves them from dissipation. From their dripping shades she distributed the waters which she garnered from the skies over the plain and the valley; and the husbandman, before he learned the marvels of science, worshipped the beneficent Naiad, who drew the waters of her fountain from heaven, and from her sanctuary in the forest showered them upon the arid glebe, and gave new verdure to the plain. After science had explained to us the law by which these supplies of moisture are preserved by the trees, the Naiad still remained a sacred theme of poetry. We would not remove the drapery of foliage that protects her fountain, nor drive her into exile by the destruction of the trees, through which she holds mysterious commerce with the skies, and preserves our fields from drought.

Evelyn says: "Innumerable are the testimonies I might produce concerning the inspiring and sacred influence of groves from the ancient poets and historians. Here the noblest raptures have been conceived; and in the walks and shades of trees poets have composed verses which have animated men to glorious and heroic actions. Here orators have made their panegyrics, historians their grave relations; and here profound philosophers have loved to pass their lives in repose and contemplation."

As man is nomadic before he is agricultural, and a maker of tents and wigwams before he is a builder of houses and temples, in like manner he is an architect and an idolater before he becomes a student of wisdom. He is a sacrificer in temples and a priest at their altars before he is a teacher of philosophy and an interpreter of nature. After the perfection of mechanical science, a higher state of mental culture succeeds, causing us to see all nature invested with beauty, and fraught with imaginative charms, adding new wonders to our views of creation and new dignity to life. Man learns now to regard trees in other relations beside their capacity to supply his physical and mechanical wants. He looks upon them as the principal ornaments of the landscape, and as the conservatories in which nature preserves certain plants and small animals and birds that will thrive only under their protection, and those insect hosts that charm the student with their beauty and excite his wonder by their mysterious instincts. Science has built an altar under the trees, and delivers thence new oracles of wisdom, teaching men how they are mysteriously wedded to the clouds, and are the instruments of their beneficence to the earth.

It is difficult to estimate how great a part of all that is cheerful and delightful in the recollections of our life is associated with trees. They are allied with the songs of morn, with the quiet of noonday, with social gatherings under the evening sky, and with the beauty and attractiveness of every season. Nowhere does nature look more lovely, or the sounds from birds and insects affect us more deeply, than under their benevolent shade. Never does the blue sky look more serene than when its dappled azure glimmers through their green trembling leaves. Their recesses, which in the early ages were the temples of religion and science, are still the favorite resorts of the studious, the scenes of sport for the active and adventurous, and the very sanctuary of peaceful seclusion for the contemplative and sorrowful.

In our early years we are charmed with the solitude of groves, with the flowers that dwell in their nooks, with the living creatures that sport among their branches, and with the birds that convey to us by their notes a share of their own indefinable happiness. Nature has made use of trees to wed our minds to the love of homely scenes, and to make us satisfied with life. How many recollections of village merry-makings, of rural sports and pastimes, of the frolics of children and of studious recreation, come to us when we sit down under some old familiar tree that stands in the open field or by the wayside! Trees are among the most poetic objects of creation. Every wood teems with legends of mythology and romance; every tree is vocal with music; and their flowers and fruits do not afford more luxury to the sense than delight to the mind. Trees have their roots in the ground; but they send up their branches toward the skies, and are so many supplicants to Heaven for blessings on the earth.

In whatever light we regard trees, they deserve attention as the fairest ornaments of nature; and the more we study them, the more do we think upon the dangers that await them from the improvidence of man. He takes but a narrow view of their importance who considers only their economical value. The painter has always made them a particular branch of his study; and the poet understands their advantages in increasing the effect of his descriptions, and considers them the blessed gifts of nature to render the earth a beautiful abode and sanctify it to our affections.

THE ALDER.

ALL persons, however ignorant of trees in general, are familiar with the common Alder. It abounds everywhere in wet places, skirting the banks of small rivers, bordering the sides of old turnpike roads, where they pass over wet grounds, filling up the basins of muddy canals, and covering with its monotonous green foliage many an unsightly tract of land, hiding and then revealing the glittering surface of sluggish stream and lonely mere. The Alder is a homely shrub, employed by Nature merely for the groundwork of her living pictures, for covering stagnant fens with verdure in company with the water-flag and the bog-rush, and as a border growth to the fenny forest, graduating its foliage by a pleasing slope down to the verdure of the plain. The assemblages of Alder constitute the plain embroidery of watercourses, and form the ground upon which many a beautiful flowering shrub is represented and rendered more interesting.

The Alder among shrubs takes the place which the grasses occupy among herbs; having no beauty of its own, but contributing to set off to advantage the beauty of other plants that flourish in the same ground. Nature likewise employs the roots of this tree as a subterranean network, to strengthen the banks of streams and defend them from the force of torrents. The Alder in New England is seldom large enough to be called a tree; it rarely stands alone, but almost invariably in clumps or larger assemblages, the different individuals of the collection forming each a single stem, almost without branches, making an outward curve a few feet from the ground, and bending inwards toward their summit.

The foliage of the Alder is homely, but not meagre, and its color is of a very agreeable tone. It is indeed a very important feature of the landscape in summer; but in autumn it remains unaffected by the general tinting of the season, and retains its verdure till the leaves fall to the ground. Nature seems to regard this tree as a plain and useful servant, not to be decked with beautiful colors or grand proportions for the admiration of the world. But, homely as it is, it bears flowers of some beauty. These consist of a profusion of purplish aments containing a mixture of gold, and hanging tremulously from their slender sprays. The extreme length and flexibility of these clusters of flowers render them exceedingly graceful, and permit them to be set in motion by the slightest breeze. The buds are seen hanging from the branches all winter, ready to burst into bloom when vivified by the first breath of spring.

THE WITCH-HAZEL.

The Witch-Hazel, or American Hamamelis, has many superficial points of resemblance to the common alder, beside its attachment to wet, muddy soils. Its ramification is peculiar; its side branches are very short, and, like the alder, it sends from one root a number of branches diverging outwards, but with an inward curvature of their extremities. The leaves are alternate and ovate, narrowest toward the stem and feather-veined. They turn to a sort of buff-color just before the flowers appear, which are yellow, having long linear petals, without beauty, growing in a cluster of four or five in the axils of the leaves. This tree is worthy of attention chiefly as a curiosity. Like the witch-elm of Great Britain, it was formerly used for divining-rods. Its magic powers might have been suggested by its remarkable habit of bearing flowers late in the autumn, thereby reversing the general order of nature; also by producing buds, flowers, and fruit in perfection at the same time. All such phenomena might be supposed to have some connection with witchcraft.

THE AILANTUS.

THE Ailantus is a native of China, where it becomes a very large tree, often attaining the height of seventy feet. It was imported into Great Britain more than a century ago, for the benefit of the silk manufacture. A species of silkworm, which was known to be hardy and capable of forming its cocoons in the English climate, is attached to this tree and feeds upon its leaves. "The *Bombyx cynthia*," says Mongredien, "thrives well in the open air (of England) in ordinary seasons, and requires no care after being once placed on the tree. About August it spins its cocoon on one of the leaflets, bending its edges inwards, so as to form a partial envelope. As the tree is deciduous, the leaf would drop and the cocoon with it, were it not that, by an instinct, the insect, before spinning its cocoon, attaches by its strongly adhesive threads the stalk of the leaf to the woody twig that sustains it. Hence the leaves that bear the cocoons are the only ones that do not drop, and there remain persistent through the whole of the winter."

This experiment with the Ailantus proved a failure; but the tree, being very stately and ornamental, continued to be cultivated in pleasure-grounds. It was introduced into the United States in the early part of this century, and is now very common in almost all the States as a wayside tree. It possesses a great deal of beauty, being surpassed by very few trees in the size and graceful sweep of its large compound leaves, that retain their brightness and their verdure after midsummer, when our native trees have become dull and tarnished.

The leaves of the Ailantus are pinnate, containing from nine to eleven leaflets, each of these being as large as the leaf of the beech-tree. It has a great superficial resemblance to the velvet sumach, both in its foliage and ramification, so that on first sight one might easily be mistaken for the other; for its branches, though more elegant, have the same peculiar twist that gives the spray of the sumach the appearance of a stag's horn. The flowers are greenish, inconspicuous, and in upright panicles, resembling those of the poison sumach. They emit a very disagreeable odor while the flowers are in perfection, impregnating the air for a week or more.

SPONTANEITY.

We are not always aware of the true sources of our pleasures, especially those agreeable sensations awakened by a view of certain kinds of landscape. I believe the sentiment of spontaneity, or our love of what seems true to nature, to be one of them; and that while the expression of this quality acts more powerfully upon men of sensitive minds, all are capable of feeling it. Spontaneity is the expression of entire freedom on the part of Nature, during the growth of plants, how much soever her free course may have been modified by circumstances previously affecting the soil and situation. Thus no less spontaneity may be seen in the wildings that cover an old fortification, or the deserted grounds of an ancient garden, than in those of a hill or a valley which has never been disturbed by man. We all admire the freedom of these growths; but we may not be aware how much they transcend in beauty the fairest works of the planter's hand.

The connoisseur of art may object to these views of the beauty of landscape, because they are based on a sentiment which is opposed to the exercise of ornamental art for its improvement. A painter, however, if he possess the soul of his art, understands that in a rural scene every building that forms part of it must either be plain and simple, or, if highly ornate, it must be very ancient. The antiquity of such a building effaces the expression of pride and pretence which, if it were new, would be painfully apparent. But the landscape-gardener's art has from its origin been so exclusively an affair of princes and nobles, that the ideas which are the foundation of the painter's art are almost unintelligible to him. He is a purveyor to the wants of a sort of rural epicurism, and of a class of men whose love of Nature would never for a moment cause them so far to forget their own personal dignity as to allow her within their grounds to wear anything but the livery of their own pride.

One of the most prominent qualities of an interesting landscape is spontaneity, consisting not only of a natural irregularity of grouping, but of such plants as are indigenous or naturalized to the soil and situation. Exotics require so much careful attention that no carelessness of arrangement could give the assemblage an appearance of unrestrained freedom. One great pattern of spontaneity is the unbroken wilderness; but this is not what we desire in landscape. Nature is greatly deficient in interest when she exhibits no connection with human life. Her original features awaken more sympathy when blended with the operations of a simple agriculture. We are pleased with those modifications of landscape which are required by the unambitious wants of man in a humble condition; while we turn with aversion from those designed only for embellishment. We would see the hills and valleys clothed, but not ornamented; for the landscape is not a garden, and in proportion as the spontaneous embroidery of nature is unimpaired by the interference of art does it affect us with pleasure.

In the wilderness, or primitive forest, vegetation is generally uniform in its growth; but in tracts which have been once reduced to tillage and then left to nature, it is always more or less grouped. We often see in this part of the country an irregular surface, consisting of hill and dale, rolling land and level meadows, once cultivated and reduced to the purposes of agriculture and then left to nature. Many such tracts have been neglected, and have lain fallow for the greater part of a century. These grounds now present a fair example of that spontaneity which is far more attractive than the tangled and uninterrupted growth of the original forest. The previous subjugation of the soil has caused the plants that have since grown up there to become beautifully grouped by tendencies which are not entirely destroyed by the labor of the rustic farmer.

As the seeds of all plants that originally occupied this tract were destroyed by tillage for many successive years, the ground must depend on seeds afterwards deposited by birds, quadrupeds, winds, and waters for the renewal of its vegetation. Wherever a cluster of thorny plants happens to obtain root, it forms a nucleus where other seeds are detained and sown. A stump of a tree or a boulder, or a heap of stones or rubbish, would constitute the centre for many similar clusters. There plants would soon spring up, and become a protection for others, which would gradually widen the assemblage, until each would become a little islet of trees and shrubs, separated by the intervening spaces of natural lawn, pastured perhaps by domestic animals, and form a style of spontaneous grouping which is entirely inimitable.

Other similar tracts, after being cleared of wood, have been left immediately to nature, before they had suffered any reduction by the plough. The renewal of the forest in such cases is always very rapid. The trees come up more closely and in greater numbers of species than on the tilled ground, but they are not grouped. The soil being full of the stumps of trees in a living state, and of the roots and seeds of many different species of plants, there is hardly a square rod in any part of the tract that is not crowded with trees and shrubs after a very few years. Every living stump of a tree gives out several suckers, that grow up rapidly into a forest of dense coppice, and the seedlings scattered among them fill up all the intermediate spaces. Of these two examples of spontaneous vegetation, the tract which has been once reduced to tillage alone presents a picturesque or beautiful appearance.

The beauty of a thing in landscape is often enhanced by being in the *wrong* place. Human hands will generally plant trees in their right places and in a proper manner, and this propriety discovers the artist. Hence the results of the rude operations of rustic laborers are picturesque, because they plant nothing for embellishment; but men of taste, while endeavoring to imitate the spontaneity of nature, produce only a ludicrous counterfeit. This remark offers but poor encouragement to art; but it shows that there are certain graces beyond the reach of art, which are nevertheless attainable by untutored hands. There is a certain absence of congruity that constitutes the charm of a spontaneous scene. Though the voluptuous eye may be delighted with a view of smooth-shaven levels, kaleidoscopic figures cut out in lawn, and the harmonic arrangements of colors in a flower-bed, we receive more passionate delight from the bramble-covered knolls, the daisied and half-obstructed footpaths, and the wild vines and trees not planted by hands, that surround the homes of laboring men in the country.

BURNING-BUSHES.

THERE is a class of plants, not all belonging to the same genus, which have received the name of Burning-Bushes from the profusion of scarlet or crimson fruit that covers their branches after the leaves have fallen. The most beautiful of these are two species of euonymus, cultivated in gardens and ornamental grounds, and bearing the names of strawberry-tree, spindle-tree, and burning-bush. The fruit is from three to five cleft, of a pale crimson, and before the leaves have dropped, which in the autumn are nearly of the same color, the tree might, at a glance, be mistaken for a bush in flames. The euonymus, though abundant in the forests of the Middle States, is not wild in any part of New England. Here it is known only as a beautiful occupant of gardens.

Another of the Burning-Bushes is the prinos, very common in wet grounds, and known in the winter by the scarlet berries, clinging, without any apparent stems, to every twig and branch, and forming one of the most attractive objects in a winter landscape. Every part of the bush is closely covered with this fruit, which is never tarnished by frost and remains upon it until the spring. This plant has never received a good specific name. It is sometimes called winter-berry, — a name as indefinite as May-flower to mark species, or human being to distinguish persons. It is also called black alder, because it has a dark rind, to distinguish it from the true alder, which is also of the same color.

The evergreen species is a more elegant shrub, with

bright green leaves of a fine lustre. It is abundant in Plymouth County in Massachusetts, around New Bedford, and in Connecticut. It it highly prized in ornamental grounds and by florists, who bind it into their bouquets and garlands of cut flowers. The leaves of this plant have some pleasant bitter properties, and were used by our predecessors as a substitute for the tea plant, under the name of Apalachian tea.

THE BUCKTHORN.

THE Buckthorn would hardly deserve mention in these pages, except that it is very generally employed for clipped hedge-rows, in the suburbs of our cities. It is a native both of Europe and America, though as it is seen only in grounds which have formerly been cultivated, or near them, it was probably introduced. It attains the height of a small tree. It is without any beauty, having a thin foliage that falls early and is never tinted. Its black shining berries are the only ornament it possesses, and its only merit is that of patiently enduring the shears of the gardener.

THE PRIVET.

THE Privet is a much handsomer shrub of an allied family. Its foliage is more delicate, both in hue and texture, not so thin, and almost evergreen. It has become extensively naturalized in our woods, and is distinguished by its clusters of white flowers in summer and its black, shining berries in autumn. It is abundant in all lands once tilled which have become wild, in the vicinity of our old towns, and was probably introduced at an early period for an ornamental hedge plant.

WOOD-SCENERY IN WINTER.

WINTER scenery has met with a remarkable share of neglect both from authors and painters. Poets have sung of winter festivals and holidays, of Christmas festivities, of garlands of holly and trailing evergreens; but they have said little in prose or verse of the beauty or the sublimity of the season's ordinary aspects. More effort has been made to divert attention from winter, as entirely disagreeable, except within doors, than to lure the mind to its attractions. Its features have been described as only waste and desolate, and what is really admirable in them has been set aside as hardly worthy of thought. It is true there is not much variety in the countenance of winter. Its expressions are wild and rude, and partake more of sublimity than beauty. It presents an insufficient number of individual objects that can be brought to the aid either of painting or poetry; so that the composition must be made up in great degree by auxiliaries drawn from the imagination.

Winter scenery is plainly monotonous. Instead of the charming mosaic of agriculture, displayed by summer and autumn in assemblages of fields, varying in color with the native hue of their different crops, we see either a dull universal waste of seared vegetation, or one broad expanse of whiteness, relieved only by the dark slender lines of fences and the broader stripes of roads and lanes winding over the face of the snow, interspersed with buildings and occasional woods and thickets. It is apparent, however, that snow increases the variety of the landscape, when it is mapped out with groves and fragments of forest, resembling wooded islets rising out of a white sea.

The charm of winter scenery is greatly heightened by the clearing of the forest, which hides the surface of the snow and causes the scene to wear less of the aspect of grandeur than of desolation. Grandeur characterizes the view wherever an almost uninterrupted expanse of some miles of surface is completely whitened with snow. The buoyancy we feel when rambling over such a landscape resembles that produced by great altitude. Our greater physical vigor in clear winter weather prepares us to be agreeably affected by surrounding views, because our thoughts are not diverted by any sense of uncomfortable exertion, as in the languid heat of summer. Our constant transition from valley to open plain, from plain to hill, and from hill to wood, keeps the mind constantly amused with new views. We are also inspired by the grandeur of the whole scene, and do not, as in summer, give ourselves up to voluptuous sensations, but to enjoyments more purely intellectual.

Our attention is not so often directed to the beauty of trees in their denuded state, as when they are dressed in foliage and adorned with flowers and fruit. But when we consider that for six months of the year all the deciduous trees, constituting the greater part of the woods, are leafless, we cannot regard their appearance at this time as an unimportant study. When trees are in leaf their primary qualities as objects in landscape are apparent; but many secondary points of beauty are almost entirely hidden under this mass of foliage. In winter, when the whole frame of the tree is exposed to view, the delicate sculpture, the forms, the angles, and the divergences of their branches, present to sight an infinite variety of picturesque appearances.

There are certain trees, however, which are almost ugly in winter, though very beautiful in their summer dress. We see nothing attractive in the horse-chestnut, the sumach, the catalpa, and the ash, in their denuded state, when the coarseness and deformity of their spray become their salient points. Of these the horse-chestnut and the catalpa are not surpassed in beauty when they are in flower, nor the sumach in its autumnal dress, nor the ash either in summer or autumn. There is as great a variety in the style of the frame and framework of different trees as in the forms and colors of their leaves and flowers. Indeed, in some respects, trees are a more interesting study in their denuded state than when dressed in foliage. In this condition single trees become more special objects of attention than assemblages. Yet it is in winter that we perceive to the best advantage the characters of a forest vista. As we pass under the interlacing branches of the trees, we observe that peculiar arch formed by the meeting and contact of those on opposite sides of an avenue. We see this appearance only in a wide avenue, where the trees have grown since it was laid out. In the pathless wood, or in a path made through the forest after the trees have attained maturity, they have no well-formed lateral branches, and display above our heads only a formless canopy

We may observe in the spray of different trees an invariable correspondence with some of their other characters. Nut-bearers, for example, have a coarser spray than small seed-bearers; trees with large or compound leaves, than those with small or simple foliage; and trees with opposite, than those with alternate branches. Hence the oak and the hickory have a coarser spray than the birch and the elm, and the large-leaved poplar than the slender-leaved willow; the ash, with compound leaves, than the maple with simple leaves, though both have opposite branches. But if a tree bears a large nut, with leaves compound and branches opposite, like the horse-chestnut, it has no spray at all. The beech-tree, however, having a very small nut, has a fine and elegant spray, not surpassed by any tree of the forest. The opposite character of the smaller branches of certain trees is never continued in the larger divisions. But the angularity of the boughs of the oak is repeated in its angular spray, and the gracefulness of the principal branches of the elm, the birch, and the lime is traced through all their minute subdivisions.

All these phenomena are interesting subjects of observation in winter wood-scenery. But the geometric beauty of the spray of trees is hardly less remarkable than its different colors. A maple wood, for example, is gray; a poplar wood is greenish olive; a wood consisting chiefly of limes, black birches, and cherry-trees has a dark shade. These differences of coloring, as seen in masses, when viewing the wood from an elevated stand, often excite the surprise of spectators; for it is only the most careful observers who have noticed this variety of shades. In many assemblages of wood that consist of an evenly promiscuous combination of species, we observe no such picturesque marks of distinction. But in all unique assemblages, of which our land affords very frequent examples, the differences between a maple, a poplar, a willow, and a lime grove are respectively very striking. The study of these shades is of considerable importance to the painter who should wish to give a true representation of a winter landscape, with reference chiefly to its wood.

Some of my most delightful wood rambles have been taken in the winter, which has always seemed to me less a season of melancholy than autumn. The sadness we feel while the leaves are falling around us and the light of noon seems but an ominous twilight passes away after these changes are completed; we resume our cheerfulness, and look forward in pleasant anticipation of spring. I have never allowed the winter to interfere with my rambling, save when the cold was intense, the weather wet or stormy, or the snow too deep for pedestrian excursions. These difficulties are seldom in the way for more than a fourth part of the season. When the snow has been hardened by repeated freezing and thawing so as to bear our footsteps, or when the ground is bare, a winter walk affords positive pleasure. At such times I have often passed a day in the woods, not only to enjoy the physical pleasure of air and exercise and the sweet odors of the pines, but also to note the changes in the face of nature, and the manners and habits of the few remaining birds and quadrupeds.

One of the most noted circumstances attending a winter ramble in the woods is their silence. But this silence is an aid to thought as well as observation, and gives importance to every sound, as the white snow gives prominence to visual objects. When the winter sun is bright and the chilly atmosphere is calm, we may listen to the distant village hum with a sensation of melody; and we catch the gurgling sounds of streams under the glistening ice, and the voices of jubilant echoes, that send back in the general stillness every sound that penetrates their secret shell. The crumpling of the hardened snow under our feet produces a tone that silence alone could turn to music; and the rustling of every zephyr seems like a living note in this solitude. The occasional voices of winter birds have a charm hardly less delightful than the melodies of June, when every note is but the part of a general chorus. In winter we listen to sounds because they are few. Even the lowing of herds is musical, reminding us that our present solitude is encompassed by life and civilization.

The wood is no longer a green recess, a temple of leafy

beauty, a sanctuary of shade, an orchestra of melodious voices. There is perhaps less solemnity within it than when it is darkened by overarching foliage. The sun shines into it and renders some little nooks more cheerful than at any other season. I have often lingered in one of these sunny retreats to watch the chickadees and woodpeckers, that never fail to appear in sight, diligently exploring every branch of the neighboring trees. It is pleasant to woo this solitude when thus enlivened by the sun, to saunter along the turfy wood-paths, still green with clumps of moss and lycopodium, to look up into the lofty trees which have parted with their shade, observing the sculptured elegance of their limbs and the intricate beauty of their spray; pondering on the rare carvings of their bark, broken into many geometrical forms, and the curious devices of nature displayed in the incrustations upon their surface.

Sometimes a solitary evergreen stands in our way, shedding upon the hoary wood some of the greenness of summer. We should know but half of what is open to observation if we never visited the forest in the winter, and we should miss one of the most remarkable features of a winter landscape if the coniferous evergreens were absent from it. Sad and sombre as they appear when the deciduous trees are putting forth their light-green leaves, they are great heighteners of the beauty of a winter scene, and are more valuable than any other woods as a protection from wind and cold.

THE LARCH.

THE Larch, though one of the coniferous trees, is not an evergreen. It is generally known in this country as the Hacmatack, a name given it by the Indians. In favorable situations it attains a great height, though we are familiar with it as a tree of but ordinary size and stature. Its branches are very numerous, and irregularly disposed at right angles with the main stem, and not in very apparent whorls. The terminal branches are small and numerous, making considerable spray, but without much character. The American and the European Larch do not differ in their manner of putting forth their larger branches, nor in their botanical characters. They are distinguished, however, by an important difference in the style of their secondary branches. The European tree has a graceful hanging spray, drooping perpendicularly from its horizontal boughs, and swinging in the wind like that of the Norway spruce. The American tree has a shorter spray, not in the least pendent, with an appearance of more sturdiness, and less formality of outline. It displays, therefore, less of that beauty which is caused by flowing lines; on the other hand, it exhibits more firmness in its general aspect, and is a more stately tree. I prefer the American Larch because it departs further from that primness which distinguishes the coniferous trees. As it increases in height, it loses its tapering summit, and forms a head of flattened and irregular shape.

The Larch bears no part in romantic history. Neither the ancient poets nor historians say much about it.

Hence it is probable that it was not abundant in the forests of the southern part of Europe in the days of Homer and Virgil. Even its importance in furnishing the most durable wood for naval purposes is a discovery of modern times, and not until a very late period was it employed as an ornamental tree. The Larch is reputed in Europe to surpass all other trees as a fertilizer of the soil by the decomposition of its foliage. Another of its advantages, when used for plantations, is its thrifty habit on lofty sites, having a more elevated range than any other tree of equal importance. Gilpin remarks of the European tree: "It claims the Alps and the Apennines for its native country, where it thrives in higher regions of the air than any other tree of its consequence is known to do, hanging over rocks and precipices which have never been visited by human feet. Often it is felled by some Alpine peasant and thrown athwart some yawning chasm, where it affords a tremendous passage from cliff to cliff, while the cataract, roaring many fathoms below, is seen only in surges of rising vapor."

The American Larch tends to uniformity of shape when young and to variety when old. Yet the fine pyramidal forms of the young trees, and the fantastic and irregular shapes of those of older growth, are equally characteristic. The foliage is of a light green with a bluish tinge, turning to a deep orange in November, just before it falls. The bright crimson cones of the Larch, that appear in June, may be reckoned among its minor beauties. This tree is more abundant in Maine and New Hampshire than in any other part of the United States, though even there it is scarce compared with other conifers. Above the St. Lawrence, however, as far as Hudson's Bay, it forms assemblages of several miles in extent.

THE HEMLOCK.

THE Hemlock is confessedly one of the most beautiful of the coniferous evergreens, though rather narrow in its dimensions. The principal branches are small and short with very slender terminations, in which it differs from all the other spruces. The multitude of these slender sprays, and their rows of soft delicate leaves, cause those beautiful undulations that characterize the foliage of this tree when moved by the wind. The leaves, of a light green on their upper surface and of a silvery whiteness beneath, are arranged in a row on each side of the branchlets. But while those of the other spruces are sessile, those of the Hemlock have slender footstalks, yielding them a slight mobility. The spangled glitter of the foliage is caused by a slightly tremulous motion of the terminal sprays.

In a deep wood the Hemlock shows some very important defects. There it forms a shaft from fifty to eighty feet in height without any diminution of its size, until near the summit, where it tapers suddenly, forming a head of foliage that projects considerably above the general level of the forest. The trunk is covered with dead branches projecting from it on all sides, causing it to wear a very unsightly appearance; and when the tree is sawed into boards, they are found to extend directly through the sapwood of the tree, making a hole in it as round as if it were bored with an auger. This is caused by the continued growth of the trunk of the tree after the decay of its branches, every year forming a new circle round the branch, but not inosculating with it, as in other trees.

The full beauty of the Hemlock is displayed on the edge of a wood, or on a plain where it has grown without impediment, feathering down to the ground. Here we observe how much less formal it is in shape than other conifers. When there are no gaps in its ramification, the numerous branches are mostly in close contact at their extremities, so that, when viewed from the outside of the wood, it seems nearly one uninterrupted mass of foliage, hiding the interior of the tree almost entirely from sight. In its perfection, when it has enjoyed an isolated growth, without any mechanical accident to mar its symmetry, it presents a fine tapering form without stiffness, and a mass of glittering foliage with which that of but few other trees is comparable.

The branches of the Hemlock are very numerous, perfectly horizontal, and remarkable for the absence of those regular whorls that distinguish other trees of this genus. They are put forth irregularly from all parts of the trunk, turning from their horizontal position gracefully upward, drooping a little at their termination, and endowed with great flexibility. The branches are minutely subdivided, forming with their leaves a flat surface, somewhat like the compound pinnate leaves of the cicuta, or poison hemlock. From this resemblance it undoubtedly obtained its name. These branches lie one above another, each bending over at its extremities upon the surface of those below, like the feathers upon the wings of a bird.

The bark of the Hemlock is of a reddish brown, divided by furrows that separate it into scales. The young trees have a smooth bark, like that of the balsam fir. The cones are very small, numerous, and pendent, of a fine crimson color when they first appear, attached to the ends of the branches, and arriving at maturity in the autumn. The Hemlock occupies all kinds of soil, though trees of a large size are found only where it is deep and fertile. It is fond of moisture, often extending its graceful boughs from the summits of granitic rocks and declivities wet with perpetual springs. "The Hemlock is natural to the coldest regions of America, and begins to appear about Hudson's Bay, near Lake St. John; in the neighborhood of Quebec it fills the forests, and in Nova Scotia, New Brunswick, the States of Maine and Vermont, and a considerable part of New Hampshire, it constitutes three fourths of the evergreen woods. Further south it is less common, and in the Middle and Southern States it is seldom seen, except on the Alleghanies."

PINE WOODS.

I HAVE often thought of the pleasure I should feel on entering a forest of tree-ferns, and observing their elegant fronds spread out above my head, displaying a form of vegetation never witnessed except in a tropical country. Yet I doubt whether an assemblage of tree-ferns, a grove of magnolias, or an island of palms could equal a forest of pines in the expression of grandeur and solemnity. A pine wood possesses characters entirely unique, and affects us with sensations which nothing else in nature seems capable of inspiring. Whether this arises from the contrast between the light outside and the darkness within, — a certain harmonious blending of cheerfulness and gloom, — or from the novelty of the whole scene, there comes up from every deep recess and shadowy arbor, every dripping dell, every mossy fountain, and every open glen throughout the wood, an indescribable charm. Notwithstanding the darkness of its interior, and the sombre character of its dense masses of evergreen foliage, as seen from without, — whence the name of *black timber*, which has been applied to it, — yet the shade and shelter it affords, and the sentiment of grandeur it inspires, cause it to be allied with the most profound and agreeable sensations.

In a pine wood Nature presents one of her most remarkable features; and there is so much that is healthful and delightful in its emanations, and in the atmosphere that is diffused around it, that she has not denied its benefits to any clime. Pines are found in every latitude save the equatorial region, where the broad-leaved palms supply the same enduring shade. Even there pines are distributed over the mountains at a height corresponding with the northern temperate zone. Nature has spread these trees widely over the earth, that the inhabitants of the sunny South and the inhospitable North may equally derive benefit from their protection and their products. There is not a region this side of the equator, where a man may not kneel down under the fragrant shade of a pine wood, and thank the Author of nature for this beneficent gift.

In New England the white pine usually predominates in our evergreen woods, mixed in greater or less degree with pitch-pine and fir. In the gracefulness of its foliage, in its lofty stature and the beautiful symmetry of its wide-spread branches, the white pine exceeds all other species. But the balsamic fragrance that is so agreeable to travellers when journeying over the sandy tracts of some parts of New England comes from the more homely pitch-pine. These odors greet our senses at all seasons, but chiefly during the prevalence of a still south-wind, and are in a different manner almost as charming as a beautiful prospect.

In a dense pine wood we observe certain peculiarities of light and shade seldom seen in a deciduous wood. The foliage that forms the canopy over our heads is so closely woven, that, wherever an opening occurs, the light pours into it with distinct outlines of shadow, very much as it shines into a dark room through a half-opened shutter. These sudden gleams of light, blending with the all-pervading shadow in which we are involved, deepen all our sensations, and cause us to feel a little of that religious awe which is inspired when passing under the interior arches of a cathedral. The presence of a group of deciduous trees always becomes apparent at some distance before we reach it, by the flickering lights among their loose foliage, and a general brightness and cheerfulness in the space occupied by the group.

There are many other agreeable circumstances connected with a pine wood. The foliage that drops from the trees, after the new growth of leaves has been put forth, covers the ground with a smooth brown matting, as comfortable to the footsteps as a gravel walk, while it savors only of nature. The acicular foliage of the pine is so hard and durable, that in summer we always find the last year's crop lying upon the ground in a state of perfect soundness, and under it that of the preceding year only partially decayed. This bed of foliage is so compact as to prevent the growth of underbrush; and it keeps the space open under the trees, whose tall shafts resemble pillars rising out of the floor of a magnificent temple. Hence a pine wood is pleasantly accessible to the rambler and the student of nature; and the absence of a woody undergrowth permits many plants of a peculiar character to thrive upon this carpeted ground. The purple cypripedia is common here, pushing up its leaves through this mass of decayed foliage, and displaying its beautiful inflated blossoms like some bright flower of a fairer clime. Mushrooms of various species and of divers fantastic shapes are frequent as we pass, some spreading out their hoods like a parasol, some with a dragon-like aspect, others perfectly globular, all having a great diversity of hues. In the deeper wood, where there is no sunshine to green the sprouting herbs, appears that rare genus of plants resembling the pale and sickly slaves of the mine, — the grotesque and singular monotropa.

In an old pine wood our attention is diverted by the great variety of lichens that incrust the bark of the trees and hang from their boughs. Many rare species decorate the trees with their tufts, circles, and protuberances, and their curiously painted dots and patches. All green herbs, however, are checked in their growth by the darkness of the wood. The verdure of a pine wood is chiefly over our heads; there is but little under our feet. But the few plants whose habits permit them to grow here are the more conspicuous because they are not mingled with a crowded assemblage of different species. Hence the little creeping michella, with its checkered green leaves, its twin flowers resembling heath-blossoms, and its scarlet fruit, is very beautiful, clustering at the roots of some tall pine, or garlanding some prostrate tree covered with mosses that mark its decay.

In the Southern States, extensive regions called "pine barrens" form a very conspicuous part of the scenery. Their growth at the present time is seldom so dense as that of a Northern pine wood. Whole forests are so thinly set that you may drive some miles through them on horseback. Still in these pine barrens there is the same breathing of solemnity that makes a Northern pine wood so impressive. The tall, gaunt, and grotesque forms of the trees, the flat, interminable plains which they occupy, the dark drapery of moss that hangs from their boughs, their silence and solitude and their primitive wildness, yield the scene an expression of melancholy grandeur that cannot be described. Occasionally a log-hut varies the prospect, as primitive in its appearance as the wood.

The pine barrens of the Southern States are celebrated as healthful retreats for the inhabitants of the seaports, whither they resort in summer to escape the prevailing fevers. They are generally of a mixed character, consisting of the Northern pitch-pine, the long-leaved pine, and a few other species, intermixed with the Southern cypress, occasional red maples, and a few other deciduous trees. Pines, however, constitute the dominant growth; but the trees are, for the most part, widely separated, so that the surface is green with herbs and grasses, and often covered with flowers. The thinness of these woods may be attributed to the practice, for two centuries past, of tapping the trees for turpentine, causing their gradual decay. Their tall forms and branchless trunks show that they obtained their principal growth in a dense wood.

The first visit I made to the pine barrens was after a long ride by railroad through the plains of North Carolina. It was night; and I often looked from the car windows into the darkness, made still more affecting by the sight of the tall pines that raised their heads almost into the clouds, like monsters watching the progress of our journey. The prospect was rendered almost invisible by the darkness that gave prominence to the dusky forms of the trees as they were pictured against the half-luminous sky. At length the day began to break, and the morning beams revealed to my sight an immense wilderness of giant spectres. The cars made a pause at this hour, allowing the passengers to step outside; and while absorbed in the contemplation of this desolate region, suddenly the loud and mellow tones of the mocking-bird came to my ears, and, as if by enchantment, reversed the character of my thoughts. The desert, no longer a solitude, inspired me with emotions of unspeakable delight. Morning never seemed so lovely as when the rising sun, with his golden beams and lengthened shadows, was greeted by this warbling salutation, as from some messenger of light who seemed to announce that Nature over all scenes has extended her beneficence, and to all regions of the earth dispenses her favors and her smiles.

At the end of my journey I took a stroll into the wood. It was in the month of June, when vegetation was in its prime, before it was seared by the summer drought. Many beautiful shrubs were conspicuous with their flowers, though the wood contained but a small proportion of shrubby undergrowth. During my botanical rambles in this wood, I was struck with the multitude of flowers in its shady arbors, seeming the more numerous to me as I had previously confined my observations to Northern woods. The phlox grew here in all its native delicacy, where it had never known the fostering hand of man. Crimson rhexias — called by the inhabitants deerweed — were distributed among the grassy knolls, like clusters of picotees. Variegated passion-flowers were conspicuous on the bare white sand that checkered the green surface, displaying their emblematic forms on their low repent vines, and reminding the wanderer in these solitudes of that faith which was founded on humility and crowned with martyrdom. Here too the spiderwort of our gardens, in a meeker form of beauty and a paler radiance, luxuriated under the protection of the wood. I observed also the predominance of luxuriant vines, indicating our near approach to the tropics, rearing themselves upon the tall and naked shafts of the trees, some, like the bignonia, in a full blaze of crimson, others, like the climbing fern, draping the trees in perennial verdure.

THE FIR.

THE Fir and the spruce are readily distinguished from the pine by their botanical characters and by those general marks which are apparent to common observers. They have shorter leaves than the pine, not arranged in fascicles, but singly and in rows along the branch. The cones of the American species are smaller than those of the pine, and they ripen their seeds every year; their lateral branches are smaller and more numerous, and are given out more horizontally. They are taller in proportion to their spread, and more regularly pyramidal in their outlines. The principal generic distinction between the Fir and the spruce is the manner in which they bear their cones; those of the Fir stand erect upon their branch, while those of the spruce are suspended from it. Botanists have lately separated the spruce from the Fir, which they describe under the generic name of Picea. As my descriptions of trees are physiognomical rather than botanical, I shall have no occasion to adopt or to reject this innovation. The spruces, however, are always described by travellers as firs. Whenever they speak of Fir woods, they include in them both the Fir and the spruce.

THE BALSAM FIR.

This tree is the American representative of the silver fir of Europe, but is inferior to it in all respects. The silver fir is one of the tallest trees on the continent of Europe, remarkable for the beauty of its form and foliage, and for the value of its timber. The American tree is inferior to it in height, in density of foliage, in longevity, and in the durability of its wood. Both trees, however, display the same general characters to observation, having a bluish-green foliage, with a silvery under surface, closely arranged upon the branches, that curve gracefully upward at the extremities. The secondary branches have the same upward curvature, never hanging down in the formal manner of the Norway spruce. There is an airiness in its appearance that is quite charming, and to a certain extent makes amends for its evident imperfections. When the Balsam Fir is young, it is very neat and pretty; but as it advances in years it becomes bald, and displays but little foliage except on the extremities of the branches. This is a remarkable defect in many of this family of trees. European writers complain of it in the silver fir. It is observed in the hemlock, except in favorable situations, and in the black spruce, but in a less degree in the white and Norway spruces.

GRANDEUR AND SUBLIMITY.

MANY of our most agreeable emotions are but modifications of a painful sensation; and in this respect there is a remarkable analogy between our mental and our physical being. While a small portion of certain substances will act agreeably and healthfully upon the organs of taste and assimilation, an excess would be offensive and perhaps fatal. Light, the source of all visual pleasure, would in a certain excess produce blindness; and circumstances that excite terror may, when combined with a consciousness of security, awaken the agreeable emotion of sublimity. Such are the effects produced by the sound of thunder at a great distance; but when it is crashing directly over our heads, the feeling of sublimity is changed to that of terror. In like manner, the intense grief we feel from the death of a friend, when partially subdued by time, becomes modified into an agreeable sentiment of reverence for the dead; and though the passion of anger is painful, that mollified anger which is termed indignation becomes pleasant by stimulating the mind with healthful resolutions.

The sentiment of grandeur seems to me to differ very considerably from that of sublimity, inasmuch as the one is a modification of agreeable and the other of painful sensations. I have remarked in another essay that a certain number of figures harmoniously arranged awaken the emotion of beauty. If these images, especially if they are brilliant, should be infinitely multiplied, their excessive multiplication exalts our sense of beauty to that of grandeur. A discernment of their disposition and properties is necessary to enable us to feel the beauty of certain harmonic figures; but the feeling of grandeur is more indefinite. A few meteors, or falling stars, are confessedly beautiful; let them be multiplied so as to cover all the visible heavens, and our sensations would be raised to the point of grandeur. But if at the same time we believed that this meteoric shower portended an immediate national calamity, the pathos mingled with the phenomenon by our superstitious fear would turn our emotion of grandeur into that of sublimity. I know it is not usual to make any considerable distinction between these two sentiments; but it seems to me perfectly compatible with the general usage of the two words to distinguish them. A certain excess of those qualities that produce a sense of physical beauty, as an infinite multiplication of beautiful lights, causes the emotion of grandeur. On the other hand, a dim discernment or sensation of the awful or the terrible causes the emotion of sublimity. We may apply the same remarks to sounds. A loud crash of harmonious and musical sounds produces a sense of grandeur; an equally loud combination of discordant sounds, so far distant as not to excite terror, awakens a sense of sublimity. Grandeur is purely exhilarating; sublimity, though certainly an agreeable sentiment, is always more or less depressing.

Sounds are more frequently a cause of the sublime than sights, because the ear is a more emotional organ than the eye. Music is therefore more easily rendered sublime than painting. In a cathedral, while the understanding is informed by the painted scenes, the passions are excited by musical strains from the choir; and the solemn grandeur of the interior becomes completely effective only when aided by the chant or the anthem. Darkness, solitude, and silence are aids to the sublime, and with their aid a still deeper effect may be produced on the imagination by spectral illusions than by sounds. And the fact that we are more powerfully affected by shadows than by substances agrees with the well-known power of obscure images in poetry, painting, and romance.

The emotion of sublimity seems to me to be more simple than that of grandeur, and more allied to solemnity. A single melancholy note in the silence of a deep forest, or in the solemn stillness and darkness of night, would cause an emotion more like that of sublimity than of grandeur; and we may note a similar distinction in written compositions. In sublime writings the language is simple; though the image conveyed to the mind be obscure, the words are plain and few. In passages of grandeur, of which we find many extraordinary examples in the works of John Ruskin, there is a rapid enumeration of striking images that produce a dazzling and bewildering effect on the reader's mind, like a vast stream of scintillations of many brilliant colors from fireworks. A sublime description is generally indebted to a single image for its effects. Poets have often embellished their descriptions with supernatural imagery. Hence the sublimity of many passages in the Old Testament and in Ossian. In almost all cases a certain amount of vagueness or obscurity increases the force of the image. None would deny the sublimity of this passage: "I heard the voice of a great multitude, as the sound of many waters, saying, Alleluia."

By some writers the profound is placed in opposition to the sublime. This is not their proper relation to each other. The opposite of the sublime is the low and the trivial; the profound is only a modification of the sentiment. Sublimity literally refers to great altitude, profoundity to great depth. We speak of a sublime height and of a profound abyss; of a sublime poet and a profound philosopher, because the one is supposed to take flight to heaven, and the other to penetrate into the deep and hidden laws of nature. It is common also to speak of a profound, rather than a sublime, mystery, as the word seems to apply to something that is buried.

The marvellous is but a modification of the sublime. It is this sentiment that causes the pleasure we feel on beholding any unusual phenomenon in the heavens. The pleasures of mystery are founded on the same instinct, and are all interwoven with our ideas of sublimity. Even the passion of love is heightened in its origin by certain mysterious incidents connected with the life and habits of the object of the passion. All the pleasures of life are enhanced by the mystery involved in the future. That ignorance is the cause of the emotion of sublimity, in many cases, will not be denied. A perfect understanding of the proximate causes of certain natural phenomena deprives them of that mystery which renders them sublime. To the superstitious, the omen that is seen in the clouds or heard in the wind, the spirit that walks in darkness like a half-illumined shadow, and the goblin that is seen to flit across the moor in the shape of a ball of flame, are each a source of sublimity. The man of science views these phenomena with very different sensations. His knowledge of their natural causes divests them of their power over his mind. But what he has lost on the one hand he has gained on the other. Though for him Science has driven the elves from the unfrequented wood, and the fairies from their moonlight haunts, she has opened to his mental vision new heavens and new earths; and he finds new and more delightful sources of sublimity in the dim region of distant worlds that lie beyond our mortal ken.

THE SPRUCE.

The Spruce, which is indigenous in New England, comprehends the White and the Black Spruce and the Hemlock. The etymology of this word is worthy of notice. Evelyn says, "For masts (speaking of firs), those from Prussia, which we call Spruce, and Norway are the best." The word seems to be a corruption of "Pruse," meaning Prussian. I have formerly thought that the name was applied to this tree to distinguish it from others of the same family which display less of this formal symmetry; but the fir proper is certainly more *spruce* in its shape than the more flowing Spruce Fir.

THE WHITE SPRUCE.

The White Spruce is less common as an ornamental tree than the Norway spruce, which is preferred as more rapid growing and stately. But the points of difference seem to me very much in favor of the White Spruce. We may distinguish them by the following marks. The White Spruce is not so tall as the European tree, and its cones are very much smaller, though both are pendent. But what is most remarkable is their different mode of branching. The principal branches of each are given out at right angles, with this apparent difference only, that the whorls are more widely separated in the Norway spruce, the distance seeming to be proportional to the comparative length or height of the trees. The leaves of the Norway spruce grow only on the top and two sides of the branch, those of the American spruce cover its whole circumference, being almost cylindrical.

But the most remarkable difference is observed in the disposition of the secondary branches. The Norway spruce suspends them almost perpendicularly from its horizontal boughs. Those of the American tree are tufted, not pendulous, but merely drooping a little at their extremities. This gives the whole mass a more sturdy appearance, and takes away some of that formality which is so tiresome in the Norway spruce. For we should bear in mind, that, although hanging foliage is supposed to be less formal than the opposite, it is not invariably so. The drooping foliage of the elm and the hemlock is graceful, but that of the Norway spruce resembles an artificial arrangement, and reminds me of garments hanging upon a patent clothes-line. I think the tufted mode of growth of the American spruces would be generally preferred to the formal drooping foliage of the Norway spruce and European larch.

THE BLACK SPRUCE.

The Black Spruce is a taller and larger tree in its native forest than the white spruce ; but the latter, when planted in pleasure-grounds, makes a more beautiful standard than the other, which is apt to grow scraggy and defective, like the balsam fir. There is some difficulty in distinguishing the two American species, until they have been repeatedly examined and compared, though they do not differ from each other so obviously as they both differ from the Norway spruce. In the white spruce the trunk tapers more rapidly, the bark of the recent branches is lighter colored, the cones are smaller and more elongated, the leaves have more of a glaucous hue, they are also

longer and less numerous, and do not form so perfect a cylinder by closely surrounding the branch, as in the Black Spruce.

Notwithstanding their similarity, it is the Black Spruce alone that produces the essential oil for the manufacture of beer. This species is also much more valuable for its timber. Emerson remarks that the leaves and scales of all the pine family, in which are included the spruce and the fir, are so disposed as to form spirals in two directions.

THE NORWAY SPRUCE.

The Norway Spruce is very favorably known in this part of the country as an ornamental tree. It is described by European writers as the tallest tree of the European forest, except the silver fir. In this country no trees of this species have attained any great altitude, having been all planted within a space of fifty years. Occasionally we behold a solitary individual that may have attained about half of its possible height, but the most do not exceed twenty or thirty feet. In certain situations no man could help admiring the beauty and majesty of these trees, when, for example, they border an extensive field, dividing it, as it were, from the roadside, as may be seen on the southern borders of the Observatory ground in Cambridge. But as a boundary for a garden or enclosure the trees of all this family are too gloomy. The Norway Spruce would be more valuable to plant for its timber than our native species, because it is more rapid in its growth and would produce a greater length of shaft in a given number of years. But the two American spruces are are more beautiful trees, as would be apparent to any one who should compare them when growing together.

RELATIONS OF TREES TO ORNAMENT.

TASTE is a very arbitrary principle, the result of certain conventional rules as changeable as fashion, and not essentially differing from it. So various are the rules which have been established at different times by the arbiters of taste, that I am persuaded, if they were entirely disregarded in our operations for improving the face of the country, and if, in the place of them, we were governed by a far-seeing and rational principle of utility, we should promote in the highest degree the beauty of landscape. Taste is but another name for fashion in every department of art; and the works of those who are governed by an exclusive regard for its canons end only in the display of pride, and create a kind of scenery that is powerless in producing agreeable impressions. The most ignorant rustic would not be so apt to spoil the aspect of his farm as an "improver." The vandalism of taste is of all kinds of vandalism the most destructive of the genuine beauty of nature. If we persuade men that "the beautiful" must be seen in all their works, they will make their grounds and their buildings excessively ornate and showy, because but few have any clear conceptions of that kind of beauty which appeals to a rational and poetic sentiment.

It will be said, on the other hand, that our ancestors disregarded all considerations of taste, governing their practice entirely by their ideas of utility and convenience; hence the bald and unsightly appearance of many old New England homesteads. I reply that these old places in general constitute the most picturesque and delightful farm scenery on this earth; but wherever the baldness attributed to them is apparent, it has been caused by avarice and narrow views of economy, guided by an entire ignorance of the value of certain important natural objects. The man who cuts down his trees and shrubbery from places where the economy of nature requires their preservation is actuated by a sense of immediate pecuniary gain, not by a rational sense of utility. The ruinous operations of some of our predecessors were caused, not by a want of taste, but by a want of knowledge. To show the truth of my assertion that a broad and far-seeing principle of utility is sufficient to guide our hand in order to produce the most beautiful and impressive kind of landscape, I will trace the operations of two of my neighbors, one a philosophic agriculturist, the other a "man of taste." I shall endeavor to make it appear that physics is a far better teacher than æsthetics, if we would learn how to beautify the face of nature.

My philosophic neighbor has never studied artistic effects in the management of his wood and ground. He operates exclusively according to his ideas of utility. He neither plants nor builds anything for ornament or the display of art. He desires only to have a convenient house and a profitable farm. For these ends he has gathered about his estate trees and shrubs of all native species, not designing them as ornaments, but as instruments for accomplishing certain valuable purposes. He comprehends the full value of all different natural objects in the economy of a farm, and takes special pains for their preservation; and he considers decorative art injurious to the simple and rustic beauty that appertains to a farm. But the neat and orderly condition of all useful objects, whether they are groups and rows of trees, or border growths of shrubbery and little shelters for birds, and that fitness and propriety which are everywhere conspicuous, combined with an evident regardlessness of display, mark the genuine farm and the intelligent husbandman.

My neighbor has a superior understanding of the laws of nature. He has studied all the complicated relations of things in Nature's economy; he sees how her benevolent designs may be carried out for our own benefit, and how, on the other hand, they may be thwarted by certain simple operations, not imagined by others to be of any importance. There is a wood on the northern boundary of his farm, standing on a gravelly hill, with a grassy slope below, which he has often been advised to cut for timber and fuel. But appreciating the advantage of a wood in this situation to protect his crops and his buildings from the northerly winds, he believes it wiser to use it as a permanent bulwark, than to fell it for its immediate value in the market. He feels assured likewise that such a barren foundation must remain ever afterwards a useless space. His sense of utility has thus been the cause of preserving one of the most beautiful ornaments of his grounds.

He occupies the level parts of his lands and the gentle slopes for tillage; but you cannot from any position on the ground obtain a clear lookout in all directions. Your prospect is interrupted by frequent growths of wood, covering little barren elevations, or projecting rocks that extend from a quarry beneath. All these prominences are covered with trees and their undergrowth, or with shrubbery and coppice. His fields, you will observe, present a very bushy appearance, more shaggy and rude than would please the "eye of taste"; for he allows two or three feet of space on each side of his fences to be filled with trees and shrubs. He has planted all vacant spaces in these borders with them, if they did not come up of their own accord, and never permits the ground beneath them to be disturbed, or any part of the undergrowth to be cleared. Two important advantages, he says, are gained by this management. The winds that sweep across his fields are checked in their force by these barriers of trees and shrubbery; so that he gains more by the protection they afford his crops than he loses by leaving so much of his land untilled. The second advantage is derived from the shelter they afford to birds, and the consequent diminution of grubs, caterpillars, and other injurious insects. But this is not all; there is no end to the pleasant walks by these natural hedge-rows, the sunny and protected nooks, the cool shady paths, to say nothing of the delightful seclusion they afford and the innumerable variety of flowers and fruits which they bear in their proper seasons.

On a certain part of his farm, encompassing a considerable space, is a steep ridge, occupying perhaps a furlong in absolute length. It is one of those moraines, popularly denominated an Indian ridge, consisting chiefly of sand and pebbles. It is too steep and narrow to admit of cultivation; the soil is barren, and the crops, if any could be raised from it, would be frequently destroyed, during showers, by the forcible descent of their waters over the unobstructed surface. My neighbor has kept this ridge covered with its native growth of trees and underbrush, removing a few trees from time to time, for fuel or lumber, but only so fast as it could be done without sensibly diminishing the quantity. Strangers visiting the place admire the hanging wood upon this declivity, overlooking the cultivated fields below, and enclosing them in its delightful umbrage. They all praise the taste of the proprietor, who has taken such pains to preserve these beautiful features of his place. He acknowledges that he is pleased with their beauty, but the visitors are surprised when he declares that he has not preserved a single object from any consideration of taste. On the other hand, he convinces them that trees are more profitable to him on this ridge than anything else that could be made to grow there. They prevent inundations from showers, protect the grounds from wind, and in a hundred ways improve the local climate and increase the productiveness of his farm.

Let me follow my philosophic neighbor in other measures that contribute to the beauty of his grounds, though designed only for some economical purpose. A small river winds through his farm, which is liable to overflow and wash the soil from its banks. To guard them from this accident, he has preserved the trees, shrubs, and vines upon them, except those parts which by their condition are exempt from danger. In one place there is a small peninsula, which is formed by a sudden bend in the course of the river, and would be liable to be demolished by an inundation, if its banks were not strengthened by nature or art. This neck of land is kept entirely covered with trees and shrubs, whose interwoven roots prevent the encroachment of the stream. The little grove thus maintained upon it offers the birds both shelter and seclusion, and in spring and summer it is a perfect orchestra of woodland song. It is delightful to sail down this river on a calm summer's day, under the shade of the trees that frequently meet and interlace their branches from the opposite sides, and view the landscape from the frequent openings. The grounds of the whole farm are picturesque and beautiful, though nothing has been done or left undone except for some useful and economical purpose.

My other neighbor is a man of æsthetic proclivities, — a lover of the "beautiful," — and, having considerable wealth, he determined to convert his estate into a *ferme ornée*. His grounds are so similar in their natural conformation to those I have described, that no important difference could be discovered. But in their present aspect, caused by the different management of the proprietors, no two places, originally so similar, could be more unlike. My æsthetic neighbor has made a grand display of the fine arts upon his estate. The grounds immediately adjoining his house are indeed a museum of Grecian mythology in marble. The topiary art has also been employed to an extreme. Clumps of shrubbery appear in the shape of domes and pyramids, and clipped hedge-rows are substituted for all artificial fences, and for their maintenance he has removed the stone-walls and their beautiful border growths of shrubbery, which are so pleasing a feature on the rustic farm.

He would tolerate none of those straight rows of trees, the charming growth of accident, by the sides of his fences. He could not bear any such formality. Having broken them up according to a modern canon of taste that condemns all straight lines, he has planted around the few trees which are allowed to remain a number of others, to form artistic groups, and to deceive Nature into the belief that they are her spontaneous creation. For, as a student of æsthetics, he has discovered that Nature plants her trees in bundles, with delightful spaces of lawn between them, and not, as the naturalist and forester have always observed, in tangled confusion. The weeds and tufts of sedge and brambles have been so thoroughly eradicated that not even a spike of panic-grass is left from which a sparrow might peck a few nutritious seeds. Not a bird could find a bush or a tussock in which he might nestle. The whole feathered tribe are banished from the grounds, while, on the adjoining farm, they assemble and sing and gladden every scene by their presence. The only wild birds here are visitors from adjoining farms; but their absence is supplied by a splendid aviary, where many foreign songsters, obtained at great price, are imprisoned, and make a noisy confusion of musical sounds, — a brilliant counterfeit of the aviary of nature on the other farm, showing that taste, like fashion, admires the counterfeit of many a thing which it contemns.

The ridges and elevations of his farm, so beautiful when covered with trees and their native embroidery, have been entirely cleared of their undergrowth, and of all the trees except a few of handsome size and proportions. The ground beneath them has been thoroughly spaded and smoothed, a crop of lawn grass has been sown there, and is kept short and velvety by a mowing-machine. Here are no rustic wood-paths; in the place of them nicely smoothed gravel-walks pursue a serpentine course in the "line of beauty" up hill and down and along the grassy plain in many tasteful circumgyrations. When a copious shower falls upon the hills of this model estate, where the surface is wholly cleared and smoothed, the water rushes down their slopes in forcible torrents, inundating all the plain below, and forming great gullies on the hillsides and heaps of sand and gravel at their base. My philosophic neighbor's farm escapes all these evils. Not even the little ground-sparrow is disturbed in her nest by the most violent thunder-showers.

If you were to pass over the grounds of my æsthetic neighbor, you would be affected everywhere with a sense that nature is subdued. In strict accordance with the rules of the "natural system" of landscape-gardening, everything has been done for the eye and the admiration, and nothing for the comfort and delight, of the visitor. While walking in his grounds, you are affected with a *quasi* feeling of grandeur. You can look upon a wide space from almost any point of view. This does not give you a sense of freedom; but the niceness and trimness of the grounds cause you a painful feeling of restraint.

Neither do you find seclusion here. There are no little cosey retreats, so frequent on the rustic farm, among the wild shrubbery, no solitary foot-paths leading you to sunny arbors frequented by birds and wild flowers, then down into thickets of roses and clematis; all here is open and smooth and bald and rounded and graded, not a single scene is beheld that does not remind you how closely allied are taste and barbarism.

THE NORTHERN CYPRESS.

THE Northern Cypress, or White Cedar, is a more stately tree than the juniper, but it is never seen by our waysides; it will thrive only in swampy soils. This is the tree that covers those extensive morasses known as cedar swamps, which are, perhaps, the best examples extant of the primitive forest. The White Cedar is not often called the Cypress in New England, and in general appearance, and especially in the style of its foliage, bears but little resemblance to the Southern Cypress; but its similarity to the juniper is very striking. It is a taller tree than the European Cypress. By some botanists it is classed with the arbor-vitæ.

This tree is not confined to inland moors, but is often found upon marshes which are overflowed by the tide of the ocean. Cedar swamps are common in all the maritime parts of the country. In many of them in New England the trees are so closely set that it is difficult to traverse them. Their wetness presents another obstacle to the traveller, except in winter, when the water is frozen, or in the driest part of summer. In these swamps there is a covering, in some parts, of bog-moss, from six inches to a foot deep, always charged with moisture, in which are embedded several half-parasitic plants, such as the white orchis. The White Cedar constitutes with the southern cypress the principal timber of the Great Dismal Swamp, and is the last tree, except the red maple, which is discovered when travelling through an extensive morass.

Michaux remarks that in the Southern swamps which are occupied by the Northern and Southern Cypress, the former "are observed to choose the centre of the swamps, and the southern cypresses the circumference." In the region of the southern cypress the cedar swamps are skirted by the tupelo and the red maple. There is but little superficial resemblance between the two cypresses. The foliage of the Northern tree is evergreen. "Each leaf," says Michaux, "is a little branch numerously subdivided, and composed of small, acute, imbricated scales, on the back of which a minute gland is discovered with the lens. In the angle of these ramifications grow the flowers, which are scarcely visible, and which produce very small rugged cones of a greenish tint, that change to bluish towards the fall, when they open to release the fine seeds."

THE SOUTHERN CYPRESS.

We have read more perhaps of the Southern Cypress than of any other American tree; but what we have read relates to some of its peculiarities, such as the stumps that grow up among the perfect trees, and of which, in the economy of nature, it is difficult to discover the advantages. We have read also of the immense gloomy swamps that are shaded by trees of this species; of the long mosses, called the "garlands of death," that hang from their branches, rendering the scene still more gloomy. But from all our reading we should not discover what is immediately apparent to our observation, when we see this tree, that it is one of the most beautiful of the forest.

The Southern Cypress is beginning to be prized here as an ornamental tree, and the few standards in the enclosures of suburban estates will convince any one that no species has been brought from the South that surpasses it in elegance and beauty. The larch, which is a favorite ornamental tree, will not compare with it, though there is some superficial resemblance between it and the American larch. They are both deciduous; and their foliage is brighter in the summer than that of other conifers. The leaves of the deciduous Cypress are of the most delicate texture, of a light green, and arranged in neat opposite rows, like those of the hemlock, on the slender terminal branches.

Michaux remarks that the banks of the Indian River, a small stream in Delaware, are the northern boundary of the deciduous Cypress. He says it occupies an area of more than fifteen hundred miles. The largest trees are found in the swamps that contain a deep, miry soil, with a surface of vegetable mould, renewed every year by floods. Some of these trees are "one hundred and twenty feet in height, and from twenty-five to forty feet in circumference at the conical base, which, at the surface of the earth, is always three or four times as large as the continued diameter of the trunk. In felling them the negroes are obliged to raise themselves upon scaffolds five or six feet from the ground. The base is usually hollow for three quarters of its bulk." The conical protuberances for which this tree is remarkable come from the roots of the largest trees, particularly of those in very wet soils. "They are," says Michaux, "commonly from eighteen to twenty-four inches in height, and sometimes from four to five feet in thickness. They are always hollow, smooth on the surface, and covered with a reddish bark like the roots, which they resemble also in the softness of their wood. They exhibit no signs of vegetation, and I have never succeeded in obtaining shoots by wounding their surface and covering them with earth. No cause can be assigned for their existence. They are peculiar to the Cypress, and begin to appear when it is twenty or twenty-five feet in height. They are made use of only by the negroes for bee-hives.

The leaves of the Cypress seem like pinnate leaves, with two rows of leaflets. Their tint is of a light and very bright green, which gives the tree a liveliness, when in full foliage, that is displayed but by few other trees. But as the foliage is deciduous, and as the branches in its native swamps are covered by long tresses of black moss, when it has shed its leaves nothing in nature can present a more gloomy appearance. In a dense wood, the foliage is very thin, giving rise to the name of the Bald Cypress, so that it is only on the outside of the forest that the tree can be considered beautiful. Its spray is of as fine a texture as the leaves. When the tree is young it is pyramidal, but the old trees are invariably flattened at the top.

The wood of this tree, though soft, is very durable, fine grained, and of a reddish color, and is extensively used for the same purposes for which the wood of the white pine is employed.

THOREAU.

Every student of nature or admirer of poetry as exemplified in life and action, who should make a visit to Walden Pond, would seek the spot which was made sacred by the two years' solitary residence of Henry D. Thoreau. Walden is known to the public chiefly by what Thoreau has written of it, and by his hermit life upon its borders. Society ought to have exclaimed against the present desecration of that hallowed spot by making it the ground for picnics, — assemblages of people who go there, not for the observation of nature, but for ice-creams and soda-water, and for repeating in the country the amusements of the city. Walden is not simply a beautiful sheet of water surrounded by a wild wood and adorned with water-lilies and pontederia; it is the scene of a few years of solitary life of a philosopher who lived according to his own maxims, of a poet who acted up to his own inspiration, a pious devotee who built his altar at the fountain of the Naiad and in the first temple of the gods.

He made his home under the trees, that he might listen at all hours to the music that fell with the dew-drops from their leaves. "This was an airy unplastered cabin, fit to entertain a travelling god, and where a goddess might trail her garments. The winds," said Thoreau, "that swept over my dwelling, were such as sweep over the ridges of mountains, bearing the broken strains or celestial parts only of terrestrial music. The morning wind forever blows; the poem of creation is uninterrupted;

but few are the ears that hear it. Olympus is but the outside of the earth everywhere." The pine shed its odors around him in all his daily employments. The squirrel dropped its nutshells, after its repast among the hickory boughs, upon the roof of his house, and the partridge led her little brood into his garden. Here he dwelt, a hermit, without the hermit's superstition; living not as a saint nor as a cynic, but as a priest and worshipper of nature.

Thoreau sought in the woods for the realization of a life which he thought possible to humanity, if men, after imbibing the knowledge of civilization, could still retain the simplicity of man in the infancy of society. He sought to carry into practice the beautiful visions which Rousseau described, but never dared to live. The beauty and sublimity of nature he wished to incorporate into his daily observations of life. He loved the ideal of nature as a poet loves a beautiful young woman whom his imagination has apotheosized. He, of all visionaries, had the courage to turn his visions into realities. He, of all poets, resolved to mould his life into a continued pastoral epic. He mingled his own fancies with the balm of the primitive wood. He would not lose the beautiful shape of those visions in the confusion of cities, but in the solitude of the forest hoped to perpetuate them in his life.

He says: "I went to the woods because I wished to live deliberately, to front only the essential facts of life, and see if I could not learn what it had to teach, and not, when I came to die, discover that I had not lived." In such passages as this he unfolds the secret of his heart. When he speaks of economy and some other practical topics, his philosophy proves itself impracticable; but when in his enthusiasm he utters one of his rhapsodies, then does the light that flashes from his own mind afford us a glimpse of the paradise that dwelt within him, whose delights he sought under the rustling leaves of the aspen and the musical moaning of the pine. "The universe," he said, "constantly and obediently answers to our conceptions; whether we travel fast or slow, the track is laid for us. Let us spend our lives in conceiving them. The poet or the artist never yet had so fair and noble a design, but some of his posterity at least could accomplish it."

He desired a life without laborious study or toil, not from indolence, which he never felt, but that he might exemplify the benevolence of nature in his own system of living. It was a sublime thought which only a poet could conceive and only a brave man could carry out. Some of the eremites of old believed that by dwelling alone, and giving themselves up to contemplation, they might gradually attain some of the perfection of the Deity. Thoreau had a grand conception of a certain simplicity of life, like that of rustic laborers, without their slavish toil, which he desired to illustrate by his own experiment. Could he have attained his end, we should have seen in his experience a signal exemplification of the ideal life of a shepherd as described by the poets. His was not the fanaticism of a religionist: it was the inspiration of a poet seeking manifestation in his walks, in his employments, and at the domestic board. If the rural gods had not forsaken the earth, they would have assembled in his hut to listen to his words, and would have sat with him in his house. His simple rustic neighbors respected him as a saint, and felt honored in his presence.

When enshrined in his own solitude, he devoted himself to the observation of everything around him. He listened to sounds as the ancient augurs listened to the oracles of Dodona, not to interpret from them a prophetic meaning, but to discover the effects they produce upon our feelings, and to remember them as sources of inspiration. No poet has equalled him in his descriptions of sounds. "I rejoice," he says, "that there are owls. Let them do the idiotic and maniacal hooting for men. It is a sound admirably suited to swamps and twilight woods which no day illustrates, suggesting a vast and undeveloped nature which men have not recognized. They represent the stark twilight and unsatisfied thoughts which all have."

Thoreau was a poet, rather than a philosopher. The luminous medium through which he saw all things appertaining to nature incapacitated him for logical reasoning. He lived upon his intuitions. His style of writing was very simple, occasionally flashing with brilliant metaphors, which he rarely used, but which always came unsought, and were not elaborately nailed to his sentences, like pictures on a wall. His satire is inimitable, and he utters his paradoxes with such an air of inspiration that you admire them in spite of their absurdity. He saw visions and described them like a prophet, but they were unintelligible to men of the world. He saw truths, but they were for the imagination, not the reason. "I would," he said, "rather sit on a pumpkin and have it all to myself, than be crowded on a velvet cushion. I would rather ride on earth in an ox-cart with a free circulation, than go to heaven in the fancy car of an excursion train and breathe a malaria all the way."

Thoreau always took an ethereal view of terrestrial landscape, as when listening to terrestrial sounds he tried to remember only the celestial strains that were blended with them. He thought perhaps we might in another state "look down on the surface of the air, and mark where a still subtler spirit sweeps over it." He speaks of a "lake of rainbow light in which for a short time he lived like a dolphin." In a similar luminousness of genius his faculties were always involved, tingeing every object of nature with its own light and hues. Those whose minds were too dull to perceive the hue of his genius did not respect him, but thought him a fanatic.

A simple excavation now marks the place where stood Thoreau's hut. It is in an open space between two sections of Walden woods. No remnants of the house are to be discovered on the spot. Not a stone marks the place which is sacred to his genius and immortalized by his works. It has not yet been desecrated by a monument such as men erect to those who have flattered their prejudices and exalted their pride, the proud distinction of worldlings after their death. Young trees of the forest have grown up from its cellar and near its foundation, and will soon convert his garden into a wood. Wild vines cover the surface of his little farm, and field flowers cluster round its embankments, tempting the visitor to pause and admire this pious retreat of a poet who sought to realize on earth the heaven of his own inspiration.

THE JUNIPER.

THE Juniper is an historical tree, and has been the subject of many interesting traditions, — supposed by the ancients to yield a shade that was injurious to human life; the emblem of faith, because its heart is always sound; the bearer of fruit regarded as a panacea for all diseases, and a magic charm which was thrown on the funeral pile to protect the spirit of the dead from evil, and bound with the leaves to propitiate the deities by their incense. It is not improbable that the superstitious notions respecting the power of its fruit to heal diseases gave origin to the use of it in the manufacture of certain alcoholic liquors; and it is a remarkable fact that universal belief in its virtues as a panacea should have attached to a plant which is now used for no important medical purpose whatever save the flavoring of gin!

The Juniper, very generally called the Red Cedar, and known in many places as the Savin, is well known to all our people, and is associated with the most rugged scenery of our coast. On all our rocky hills which have been stripped of their original growth the Juniper springs up as if it found there a soil congenial to its wants. On the contrary, the soil is very poorly adapted to it, for the tree never attains a good size in these situations. Its presence there may be attributed to the birds that feed in winter upon its fruit, and scatter its seeds while in quest of dormant insects among the sods. As we journey southward, we find this tree in perfection in New Jersey and Maryland; and in all the Atlantic States south of

Long Island Sound the Junipers are large and thrifty trees.

On our barren hills, near the coast, where they are so common as to be the most conspicuous feature of certain regions, they display a great variety of shapes and grotesque peculiarities of outline. Yet the normal shape of this tree is a perfect spire. When it presents this form, it is, in the true sense of the word, a beautiful object. Even its rusty-green foliage gives variety to the hues of the landscape, and heightens by contrast the verdure of other trees. This effect is the more remarkable at midsummer, when the green of the different trees has become nearly uniform in its shades. At this time the mixture of the duller tints of the Juniper is very agreeable.

The Juniper is very full of branches, irregularly disposed at a small angle with the trunk, forming an exceedingly dense mass of foliage. A singular habit of this tree is that of producing tufts of branches with foliage resembling that of the prostrate Juniper, as if a branch of that shrub had been ingrafted upon it. The berries, which are abundant in the fertile trees, are of a light bluish color, and afford a winter repast to many species of birds, particularly the waxwing. The branches, when their extremities are brought into contact with the soil, readily take root. Hence we sometimes find a clump of small trees gathered like children around the parent tree.

The trunk of this tree diminishes so rapidly in size as to lose its value for many purposes to which the wood is adapted; but this rapid diminution in diameter is one of its picturesque properties, and the cause in part of that spiry form which is so much admired in this tree. The lateral branches, always inserted obliquely, diminish in size proportionally with the decrease of the trunk. The Juniper is first discovered on Cedar Island in Lake Champlain, and, south of this latitude, extends all along the coast to the Cape of Florida, and along the shores of the Gulf of Mexico.

THE ARBOR-VITÆ.

THE American Arbor-Vitæ is a small tree growing very much in the spiry form of the juniper, but narrower in the lower part. It is like the juniper also in its numerous and irregularly disposed branches. It is not seen in the woods near Boston; and it is rare even in cultivated grounds, where the Siberian Arbor-Vitæ, on account of its superior foliage, is preferred. The American tree grows abundantly in high northern latitudes. It is remarkable, with its kindred species, for the flattened shape of its leaves; and in its native woods it is hardly ever without a mixture of yellow and faded leaves interspersed with the green and healthy foliage. The terminal branch invested by the leaflets — resembling scales, and not a true leaf — constitutes this fanlike appendage, resembling the frond of a fern. The leaves have the flavor and odor of tansy.

In Maine the Arbor-Vitæ, next to the black spruce and hemlock, is more frequent than any other of the evergreens. It delights in cold, damp soils, and abounds on the rocky shores of streams and lakes. It sometimes constitutes a forest of several acres, with but a slight intermixture of other trees, predominating in proportion to the wetness of the soil. In the driest parts of these bogs we find the black spruce, the hemlock, the red birch, and, rarely, a few white pines.

THE YEW.

In Great Britain the Yew is one of the most celebrated of trees, the one that is generally consecrated to burial-grounds, and that most frequently overshadows the graves of the dead. It is a tree of second magnitude, and remarkable for its longevity. The American Yew is seldom anything more than a prostrate shrub, resembling branches of fir spreading over the ground. It is said, however, that although it is a creeping shrub on the Atlantic coast, it becomes a tree on the coast of the Pacific; in like manner the alder, which is a shrub here, becomes a tree in Oregon and California.

In New England, the Yew is a solitary tree, growing among deciduous trees as if it required their protection. It never constitutes a forest either here or in Europe. It seems to love the shade, and when it is not under the protection of trees, it is found on the shady sides of hills, and in moist, clayey soils, but never on sandy plains. I shall not speak of the romantic customs associated with the European Yew; but the absence of this tree deprives us of a very romantic feature in landscape.

RURAL LIFE IN NEW ENGLAND.

It has often seemed to me, that, while the world is progressing in the mechanic arts and in the refinements of civilized life, we are losing ground in that healthful simplicity that marked the habits of our ancestors. Epicures taught that the secret of happiness is to preserve our tranquillity. Practical philosophy has not discovered anything in contradiction of this maxim, though the instincts of men prompt them to seek excitement. There is a kind of sentimental yearning for the quiet and simplicity of rural life, but there is a more active impulse in the breast of the young, that draws them away from humble pursuits, and forces them into the march of ambition and fortune-hunting. The wisest are those who content themselves with simple rustic occupations, without falling into habits of indolence and apathy. He is a happy man who can preserve his calmness and self-possession without losing his energy; who can sit in councils of state, and not be carried away by party zeal and ambition, and, on the other hand, can swing a scythe or hold a plough without entirely discarding more thoughtful employments. The great are they who are not controlled by those circumstances that give to other men their principal hues of character.

The life of a farmer has been a theme for the praises of poets and orators from the earliest ages. The pleasures and comforts attending his labors have been so often eulogized, that the praise bestowed upon them has come to be considered one of the platitudes of ordinary eloquence, and without any substantial truth. Indeed, I believe our people have generally set it aside as a sentimental fiction, and regard the worship of Plutus as far more rational than the quiet service of the rustic Pan. Yet of all occupations that of a farmer seems to me the most delightful and the most promotive of health and happiness. Men will not accept this theory, because they cannot divest their minds of a prejudice, which has become in the American mind almost an instinct, that no man can be happy who does not feel that he is rapidly growing rich.

The farmer, above all other men, enjoys daily intercourse with pleasant rural objects. The feast of the gods is constantly before him; and though he may seem indifferent to its pleasures, his happiness is materially improved by it. Though he may profess to care nothing for the songs of birds, or the beauty of trees and flowers, he derives from them more enjoyment than he is capable of estimating. He may not know how much happiness he owes to robust health, to active exercise, to fresh air and bright sunshine; but he does not, on account of his ignorance, derive any less benefit or less pleasure from the air he breathes, from the health that renders him buoyant and cheerful, and from the sunshine that warms and enlightens him; neither does the man who gives no heed to the songs of the birds and the beauties of landscape consequently derive no pleasure from these objects when they surround his home and mingle with his pursuits.

The simplicity of a farmer's life is one of its principal charms. His pleasures and his toils are equally rational and delightful. He goes out to freedom under an open sky, and he returns to a home unincumbered with fashion and absurd conventionalities. There are duties to be performed at certain hours; but there is seldom a day which he may not appropriate to other purposes, and enjoy it in such a way as he chooses. The machinery of nature does not require the constant watching and minute punctuality which are required by the operations of trade or manufactures. Though governed by the seasons, he is not obliged to regulate his movements by a clock.

From my earliest days I have coveted the life of a tiller of the soil; but fortune has opposed the fulfilment of my desires. Still, whenever I see men employed in the labors of the field, I look upon them as realizing in these occupations, and in the circumstances attending them, the greatest amount of happiness that can be enjoyed in this world. But the farmer must be a man of simplicity in his desires, his ambition, and his habits. He cannot be happy if he covets a life of luxury and inglorious ease, nor if he is ambitious for grandeur in the style of his house and of his grounds. It is a vain ambition that leads any man to strive after honors that do not appertain to his situation in life.

Most of the inhabitants of our cities can feel the pleasures attending a short residence in the country during the summer and autumn. How happy are they who are all the year surrounded by these enjoyments and by the pleasures that flow from this close communication with nature. In the morning they are awakened to the labors of the day by the songs of the earliest birds; at the day's close they are soothed of their weariness by a renewal of the same songs. The farmer sees the pleasures of all the seasons spread out before him. The citizen goes to market and feels thankful that his money will buy a small portion of those fruits which the farmer receives almost as the gratuitous bounties of nature. Though he may not cultivate the choice fruits of the market, the wild fruits are all around him. Raspberries in rustic profusion present him a feast without price. Whortleberries hang their black and azure clusters around every rustic by-way, and blackberries sparkle on the greensward and garland the gray rocks as if some unseen hand had been employed in rearing them. All his grounds at this season are a variegated tapestry, on which are embossed flowers that have life, and fruits that do not mock the eye with a vain semblance of beauty.

To the New England farmer what successions of glad employments lead up the seasons; to him especially who has not forsaken the simplicity of habits and the frugality that distinguished his ancestors. With a joyful heart he walks abroad, and surveys the rewards of his industry in the market-places, and beholds with pride the importance of his services to the great world. Go where we will, the fields, the barns, the granaries, the stores, and the markets are full of the immediate products of the farm. In all places are we reminded of the reaping-hook and the wheaten sheaf, and behold those gifts that teach all men to venerate the plough.

The farmer is no stranger to toil; but he regards it as one of the necessary conditions by which he may secure the united blessings of health, cheerfulness, and competence. No man so frequently beholds the rising sun, nor brushes the dew from the greensward, nor breathes the morning air still sweet with emanations from opening flowers. Whether he is holding a plough or pitching sheaves, driving his team afield or trading for the sale of his products, he is the best living example of freedom and independence.

All the seasons delight in showering their gifts upon him. They strew his path with flowers, even though he may not heed them; they fill the air with music, though he may not listen. They are not niggard of their favors, and wait for no sacrifice before they yield him their rewards. Beauty garlands all his fields with her wreaths, and prefers to linger with the rustic who does not invite her, rather than with him who, spurred by emulation, tampers with those lovely objects with which nature will not allow the impertinent interference of taste and art. Health waits upon his steps, though he neglects her laws; Fortune rewards him, though he lays no gifts upon her altar; he thrives though he has but little calculation, and his industry obtains higher reward than his neighbors' ingenuity

But why should this often uncultivated clown be so particular an object of Nature's favors? Why should Providence bless him with more content than the wealthy tradesman, who would scorn his whole possessions? Because the gifts of Nature and the blessings of Heaven are for the meek and humble, and Virtue loves to bestow her rewards upon those who are simple in their habits and frugal in their desires. Hence, though he possesses but little of the ideal, the lovers of the poetic and picturesque are delighted with his creations; and though destitute of culture, the refined and the educated are pleased with his company and remember his sayings.

We overlook in his conversation those modes of speech and those refinements of thought which we look for in circles of superior social rank. We link him by unavoidable association with the trees, the rocks, the animals, and the rude implements which are subject to his command. We detect interesting likenesses between him — with his robust frame, his swarthy countenance, and his plain language — and the shepherd swains of the pastoral poets. Bright summer suns and keen winter blasts have imbued him with a kindred cheerfulness and ruggedness; but open skies and independent labor have given him a freedom of deportment and dignity, which, with all his defects of education, elevate him above the level of a clown.

When toil ceases, the farmer's daughter covers the table with brown damask, and prepares a simple entertainment as neat as it is bountiful, and the more grateful because it is unostentatious. The laborers, with a joyful appreciation of the luxuries of the frugal board, express their delight by complimenting in their awkward way the neat hands of the damsel who has ministered to their wants. No fairy queen could move with more grace or act with more benevolence than this young girl, among these less refined swains, administering the bounties of the table. With hands as white as her teacups, and cheeks that rival the downy carmine of the peach, no lordly assemblage was ever attended by a more lively embodiment of grace, tenderness, and wit.

To the field, before they have consumed their remaining leisure in gossip and jovialty, they are summoned by the distant muttering of a dark cloud that moves slowly up the horizon. A current of wind bears it steadily in a direction that may save them from its force; but they know it may be leading along another cloud to which it has been wedded by electricity. The trembling mirage that has all the noonday been suspended over the arid fields still glimmers in the hot beams of the sun. The balm of a whole season of flowers and sweet-scented herbs is ascending like an incense-offering to heaven. A fresh breeze, laden with the moisture of the south-wind, improves all these odors, and renders toil less wearisome to the swain. All hands are now busy; the lumbering wagon has entered the field; the hay is raked into swaths and rolled into heaps, and hands and feet, rakes and pitchforks, are engaged in complicated manœuvres, by which the hay is rapidly transferred from the field to the cart, and from the cart to the hay-loft. Their implements are laid aside, and the haymakers' work is done.

The cloud that gleamed with occasional fire and stimulated the action of the haymakers by its distant and sullen roar has passed away; but it was the leader of a gloomy cohort of angry masses, slowly moving upward from the gathering-place of the storm. A dense mass of sombre cloud is rising from the west, and above it are many domes and pyramids, tipped with fire of a dazzling whiteness. Below is darkness almost impenetrable, through which we perceive the fantastic course of the lightning, as if a sudden rent in the cloud had revealed through its crevices some dazzling fires beyond. The weary swains, assembled under the roof of the cottage, sit at open doors and windows, and watch the movements of the storm, yielding themselves voluptuously to the cooler breezes that precede it. The loudening notes from the cloudy band denote its nearness. The sun is obscured, the air is darkened, the leaves whirl fitfully, and the trees bow their heads to the invisible force of the wind. Smoking rain, and flashing darkness, and crashing bolts of fire, give them an awful announcement of the beauty and sublimity of the tempest that has long threatened the land.

Hay-time passes with summer, and autumn invites the swain to a new round of toil and pastime. He rejoices in an abundance hardly to be purchased with gold by those who congregate in towns; not displayed like a shopman's wares in precise arrangements to lure the eyes of passengers, but scattered far and wide, as if they were flung gratuitously from the skies. Grapes in purple clusters, basking in the sunshine, garland the stone-wall, which seems like a natural trellis. Apples are reddening in the orchard trees, under the ripening influence of the sun, or lie in heaps of variegated colors upon the ground. Peaches with downy cheeks, wearing the blush of mellow ripeness, are drooping voluptuously from their slender boughs. Quince-trees in gleaming rows along the fences tempt the visitor with the golden apples of the

Hesperides. Every wayside in the country is adorned with a similar profusion; and glittering varieties of fruits hang from thousands of boughs and sprinkle the green turf of every orchard.

Nature has benevolently made all these objects delightful to man, that he may be tempted to join the husbandman in those pursuits which, being the most noble occupations, were assigned, according to the narrative of Moses, to the great progenitor of our race. Man, urged by ambition, leaves these peaceful avocations, to explore the seas or join the game of fortune in the city. But never in his solitary moments does his bosom cease to yearn for his ancient rural home, the rustic employments of his youth, his wanderings among the hills, his angling by green pebbly watercourses, and the innumerable pleasures of the field and farm.

Fortunate men, whose ambition, if it has sometimes led you to wander into other paths, is mainly satisfied with performing well the duties of your pleasant, laborious life! Under the shade of the elm that guards the enclosures of the old farm-house you may discourse upon the fate of nations; but the Arcadian sun never shone upon a people so blessed with all the arts that will make life happy. He is the true model farmer, though he may own some fields where a bird might find shelter in a hazel copse, or where a child might linger to pluck a wild flower, who can exhibit in his own humble and rural home the best examples of domestic virtue and happiness. If the profits of his farm are small, the wants of his family are rational and few. He lives in an unpretending house, not embellished to suit the vulgar demands of taste, but adapted to utility and convenience. But "there is an angel within the house"; and I often think, as I cast my eyes about the rooms, and look upon the plain and modest furniture, the clean and bright utensils, the ruddy and smiling children, and the neat and comfortable aspect of everything within doors and without, that here is true philosophy, where health is not sacrificed to wealth, nor simplicity to pride, nor beauty to ornament, nor charity to selfishness, nor domestic peace to fashion, nor heaven to the world. If the inhabitants of heaven ever visit the earth, this is the home they would delight to enter during their sojournment.

When gilded by the morning sun, the smoke of its chimneys gracefully curling as if to meet the silvery clouds that move quietly across the sky, its windows looking out modestly under a canopy of elms, he must have a soul that is insensible to all moral beauty who is not warmed with enthusiasm, while gazing on this picture of homely happiness, with all the goodness that dwells within it and the simple luxuries that are spread around it. Come when we may, within and without is always something to exhilarate the soul and to warm the heart in this home of peace; on a summer morning, resonant through all its leafy umbrage with the notes of birds, at noonday alive with the busy return of laborers from the field and children from school, and in the evening vocal with the sounds of music and merriment.

A group of trees at a small distance from the house, upon a slight eminence, half shades a primitive well, from which the water is raised by means of a crosspole. Similar wells were formerly numerous in all parts of the country, and wherever they are now seen they are sure evidence of ancient manners and hospitality within the house. Into this well, through the interwoven branches of elms and maples, the sun sheds upon the glassy surface a picture of gold, emerald, and sapphire, as the colors and forms of the leaves, the blue sky, and the snow-white clouds are cast upon it. We look upon this picture through a circle of embroidered mosses, ferns, and lichens, that cluster upon the old rocks, resembling a flight of rustic steps down to the fountain below.

When the farmer's daughter steps out, with her sisters, under the shade of trees or upon the green slope that fronts the homestead, no princess was ever more devoutly attended by all the lovely ministrants of nature. The gales shed around her path the incense of roses and honeysuckles, twining over the rustic porch, of lilies in the garden, and lilacs in the nook of the enclosure. Wild flowers, undaunted by the simple art that prevails in these rustic grounds, creep boldly up to the very doorstep, stealing away from the wild growth that fringes the wayside. And she who treads these paths may not envy the princess who watches the streams from marble fountains, and breathes the incense of the tropics under the roof of a palace.

THE WHITE PINE.

THE pines in general have not the formality that distinguishes the fir and the spruce. They seldom display so much of a pyramidal shape as we observe in a symmetrical fir. Their leaves are longer, and their branches not so regularly given out in whorls. They are also more generally round-headed when old; their leaves are in small fascicles, containing from two to five, while those of the fir are arranged singly along the branch or round it. The pine contains a greater quantity of turpentine than any other family of resinous trees, and many of the species are of the highest value in the mechanic arts. In the New England States three species only are known, and of these two only are common.

The most remarkable of this family of trees, and the one that comes nearest the fir in symmetry and formality, is the White Pine. But though like the fir in symmetry, it resembles it the least in all other qualities, having the most flexibility of foliage of all the pines, and bearing its leaves in fives. The White Pine, according to Michaux, "is the loftiest and most valuable of the productions of the North American forest. Its summit is seen at an immense distance, aspiring to heaven, far above the heads of the surrounding trees."

At first sight of a full-grown and well-proportioned White Pine we are struck with its evident adaptedness to all purposes of shade and shelter, in its wide-spread, horizontal branches, and in its silken tufted foliage. It is not impenetrable to sunshine, but admits it in constant flickering beams of light; and we perceive immediately that there is no other tree in whose shade it would be more agreeable to recline on a hot summer's day, or under whose protection we might obtain a greater amount of comfort in winter. The uniform arrangement of its branches in whorls, forming a series of stages one above another, its tasselled foliage in long, silky tufts at the ends of the branches, and its symmetrical outline, constitute in the most obvious sense a beautiful tree. These tufts, though not pendulous, have none of the stiff bristling appearance of the other pines; and their verdure is of a sober, not a sombre tint, though rather dull in lustre.

The symmetry or formality which some writers condemn in the style of this tree is not of a disagreeable kind, like that of the Norway spruce. It is combined both with majesty and grace, and increases the grandeur of its appearance, like the architectural proportions of a temple in which grandeur could not be produced without symmetry. This tree has much of the amplitude so remarkable in the cedar of Lebanon. Hence the look of primness, which the firs always retain, is counteracted by its nobleness and altitude. It is combined also with a certain negligent habit of its leafy robes, that softens its dignity into grace, and causes it to wear its honors like one who feels no constraint under their burden.

The White Pine has no legendary history. Being an American tree, it is celebrated neither in poetry nor romance. It is associated with no classical images, like the oak, nor with sacred literature, like the cedar of Lebanon. It has no poetic history and no reputation save what it may have derived from the easy motion of its foliage, the gentle sweep of its smaller branches, its terebinthine odors, and its pleasant, romantic shade. It has no factitious charms, but depends on its own intrinsic merits for the pleasure it affords either the sight or the mind. In New

England, the White Pine contributes more than any other evergreen to give character to our scenery. It is seen both in large and small assemblages and in clumps, but not often as a solitary standard. We see it in our journeys projecting over eminences that are encircled by old roads, shading the traveller from the sun and protecting him from the wind. We have sat under its fragrant shade, in our pedestrian tours, when weary with heat and exercise we sought its coolness, and blessed it as one of the guardian deities of the wood. We are familiar with it in all pleasant, solitary places; and in our evening rambles we have listened underneath its boughs to the notes of the green warbler, who selects it for his abode, and has caught a plaintive tone from the winds that sweep through its long sibilant leaves.

The White Pine is a tree that harmonizes with all situations, rude or cultivated, level or abrupt. On the side of a hill it adds grandeur to the declivity, and yields a sweeter look of tranquillity to the green pastoral meadow. It gives a darker frown to the projecting cliff, and a more awful uncertainty to the mountain pass or the craggy ravine. Over desolate scenery it spreads a cheerfulness that detracts nothing from its power over the imagination, while it relieves it of its terrors, by presenting a green bulwark of defence against the wind and the storm. Nothing can be more picturesque in scenery than the occasional groups of White Pines on the bald hills of our New England coast, elsewhere too often a dreary waste of homely bush and brier.

Such are its picturesque characters. It may also be regarded as a true symbol of benevolence. Under its outspread roof, numerous small animals, nestling in the bed of dry leaves that cover the ground, find shelter and repose. The squirrel feeds upon the kernels obtained from its cones; the hare browses upon the trefoil and the spicy foliage of the hypericum, which are protected in its shade, and the fawn reposes on its brown couch of leaves, unmolested by the outer tempest. From its green arbors the quails are often roused in midwinter, where they feed upon the berries of the michella and the spicy wintergreen. Nature, indeed, seems to have designed this tree to protect her living creatures both in summer and winter.

The geographical limits of the White Pine are not very extensive. It is confined to northern regions, but does not extend so far north as the red pine or the fir. In the Southern and Middle States it is seen only in the Alleghany range; but it constitutes the principal timber of the pine forests of Canada and the New England States, which Loudon says are "the most extensive in the world." The *débris* of granite affords the best soil for the coniferous trees, but the White Pine is seldom found in marshes. The tree that bears the nearest resemblance to it is the Lambert pine of California, to which our tree approximates in size. Michaux measured two trunks near the banks of the Kennebec, one of which was one hundred and fifty-four feet in length, and fifty-four inches in diameter; the other, one hundred and forty-two feet in length, and forty-four inches in diameter.

AGRICULTURAL PROGRESS.

Dr. Franklin, on seeing a fly escape from a bottle in which for a long period it had been confined in a torpid state, expressed a wish that he could be corked up in the same manner for a century or more, and then awake, like the fly, to witness the progress that had been made in his beloved country. But when I consider the inevitable tendency of steam-power to concentrate wealth into the hands of capitalists, I feel as if I should be reluctant to wake up some ages hence to view my country when the world is finished. Though steam in its application to travelling and manufactures has conferred great apparent benefits on mankind, we have reason to dread the ultimate consequences to small independent farmers of the introduction of steam-power into the operations of agriculture. However expedient the system of associated capital may be for the growth of manufactures, it would be destructive to the prosperity of small farming. The corporations, executing all their heavy labor by steam-power and by mammoth implements, would crowd out of the ranks of agriculture all whose farms were of such small extent that steam could not be profitably used by them. In competing with the companies, the small farmer would find himself in the situation of the hand-spinner and the hand-weaver who should undertake to compete with the manufactories of Lowell and Lawrence.

The system of steam-farming would make it necessary that agriculture should be carried on by large associations of capital and on a magnificent scale of operations. All agricultural implements which are moved by steam must be profitable in a certain ratio to the extent of even and uninterrupted surface which is to be tilled. On small fields it would be impossible to use them with success. Hence the necessity of farming by associated capital, and of greatly increasing the size of farms by converting many into one. Under such "improved" conditions, the present system of farming could not stand in competition with steam-farming. The agricultural corporations, with their implements operated by steam, would cultivate ten acres with less expense than is now employed in cultivating one acre. If the moral and physical improvement of mankind were to be the effect of this new system, the prospect would be delightful. But no such happy results would spring from it; laboring men, instead of being elevated into lords, would be degraded into mere machines.

Men are too prone to base their theories of human progress on the assumption that labor is a curse, and not, as it is, when freely and justly rewarded, a blessing. But labor ceases to be free when the laborers are under the control and in the power of mammoth associations. Labor then becomes servitude, which is closely allied to slavery. No one would say that, under the present circumstances of the country, the operatives in our factories, however well paid, are as free as our farmers, masons, and carpenters. When labor is performed by powerful machines, man becomes a slave to the machinery; when, on the other hand, the implements in use are small, the machinery is the servant of man. The production will be greater in the former case; but the health and freedom of man are sacrificed to obtain it. The object of the statesman and the philanthropist should be to make the people free, virtuous, and happy; and any increase of the national wealth which is obtained at the expense of the moral and physical welfare of the people is not to be desired. But it may be asked by some jealous friend of progress, if it is right to refuse to agriculture those aids which have built up our manufactures? I reply, that we should refuse to agriculture any aid which is not beneficial to the agriculturist; for the farmer is of more importance than his crops. Let us not increase the products of labor by any means that will degrade man.

To illustrate the consequences of this system of agricultural progress, we will apply it to an imagined case. We will suppose that in some indefinite period of the future, when steam-farming by associated capital has become nearly universal, there remains in a certain part of the country one of those farming villages which are now so common in our happy land. The farmers in this place are intelligent workingmen and small land proprietors, who have but little wealth except their houses, lands, and stock, and support themselves by industry and honest trade. After steam-ploughs, steam-rakes, steam-mowing-machines, and other magnificent improvements of the same kind, have swept over the country, they arrive lastly at this antiquated village, where labor is free, and where the farmers are so far behind the times as to own the lands they till, and carry on farming as in the present age of political and social equality.

These industrious farmers have ascertained by bitter experience that by using hand-implements and horse and cattle power in the operations of farming, they cannot compete with the great agricultural corporations. The agent of a new company, chartered with a hundred millions of capital, offers to these unhappy men a price for their farms, which, though far less than their original value, they feel obliged to accept, especially as a promise accompanies the offer that they shall be employed as laborers on the soil, under the direction of the officers of the company, educated at an agricultural college. The majority consent to the sale, and the remainder are forced to consent by a law of the Legislature, placing it in the power of corporations, "established for the public good," to seize upon any refractory individual's estate, after paying him what a body of commissioners have declared an equivalent. These mammoth agricultural companies, by means of political intriguing and bribery, would easily obtain sufficient influence over any legislative body to cause the enactment of such a law. This any one will believe who has had political experience, and who knows how easily the most tyrannical and unjust measures may be carried by making them party tests.

Let us now observe the consequences, after this little village of happy and independent laborers has been converted into a mammoth farm, owned by a company and carried on by steam-power. At the beginning, all the pleasant old farm-houses are removed, because they stand in the way of tillage, which is performed as much as possible in large undivided lots. All fences and boundaries, except those by the roadside, are for the same reason taken down, to open all the small fields into one; for it has been ascertained that no single field can be worked with the best advantage, unless it contains at least five hundred acres. The larger the field, the more profitably can it be worked by steam. Hence the preliminaries for steam-farming are necessarily a work of devastation. Many delightful groups of trees and shrubbery, some that skirted a winding brook, others that bordered the fences, including many standard oaks and elms, are swept to the ground, rooted up by some giant infernal machine as easily as a farmer pulls up weeds. All abruptly swelling ridges and other eminences, the charm of many a landscape, some of them beautifully crowned with trees and shrubs, and others fringed with wild flowers and covered with green herbage, and forming numerous little valleys smiling in sunshine, or sweetly sleeping under the summer shade of trees, where the flocks found a comfortable resort in all weathers, are now converted into one vast level.

The brooks are conducted into canals, and carried along in straight courses for the convenience of labor and purposes of irrigation; for it is necessary that their circuities should not interfere with the progress of the steam-plough. In fine, the pleasing variety of surface that beautified the landscape when it was in the possession of the inhabitants; those quiet, rustic lanes, fringed with wild roses, hawthorns, and viburnums, conducting from the dwelling-houses to the adjoining fields and woods; the comfortable enclosures that resounded with the lowing of cattle and the cheerful noise of poultry; and, worst fate of all, the old farm-house, where the patriarch of a small estate presided over a happy family, happy because they were free and healthfully employed, — all, all are swept away by this besom of improvement.

And where are the inhabitants? The sturdy yeoman who, though doomed to hard labor, found this labor sweet because it was voluntary, the happy and independent swain, who called no man master, and who was really a king in his own acres, is now a hired servant of the corporation. The farmers, their wives, and their children, have all been reduced to servitude in this grand manufactory of corn and vegetables. The tiller of the soil has become a slave to his occupation. Each thousand acres devoted to a single crop is managed by an agent imported from the city, who receives a large salary as superintendent, and pays out their weekly pittance to the farm laborers. In order to facilitate operations, there is a minute division of labor, as in the cotton and woollen factories. Some of the farmers are employed exclusively as shovellers; some are only drivers of cattle; some ride on the engine; all, indeed, are confined to one special manipulation.

The several families, except those who left their hereditary employment and emigrated to some other place, are tenants of wooden boxes put up close to the road for the economizing of land. All these are in exact uniformity, — "model tenements," — stuck up into the air, without garden or enclosure, and owned by the corporation. The majority of the farmers, flattered with the hope of securing their wealth, invested all their money in the corporation stock, which they were soon induced to sell at immense sacrifice, because the extravagance and dishonesty of the company's agents absorbed all their profits and cut down their dividends. In less than ten years every one of these independent farmers was a poor man; and the village children, who lived as free as the birds of the air in their former rural homes, now work in platoons upon such parts of farm labor as they are able to perform. Before the village was sold and converted into a mammoth farm, you might see the little children with their satchels going regularly to the district schools, clad in neat and various attire, skipping and playing on the road, full of gladness and freedom. Now they are called up in the morning by the ringing of a bell. They rise, they work, they eat, they go to bed, and they sleep to the sound of a bell that tolls dismally in their weary ears the knell of all their former joys.

In the story of this once happy village and its inhabitants we may read the fate of the whole country if the steam-engine should ever be introduced into the operations of farming, which would, as an inevitable consequence, be carried on by associated capital. Such a class as our independent laboring farmers — the only undegenerated class in any civilized country — would cease to exist. If it be progress or improvement to convert all these hardy swains into hirelings under the agents of wealthy institutions, then progress is a curse and improvement a process of degeneracy. I am unwilling to admit any measures to be progressive that lessen the happiness and liberty of men, how much soever they may increase the national wealth or the productiveness of the arts.

THE PITCH PINE.

The Pitch Pine differs very widely in its style of growth from the white pine, and displays fewer of those points that excite our admiration. Its leaves form larger and more diffusive tufts, and are more bristling and erect from their superior rigidity. It is remarkable for its rough and shaggy appearance; hence its Latin name, *rigida*. Indeed there is not a tree in our forest that equals it in the roughness that is manifest in every part of it and in every stage of its growth. This is one of the most common trees in the Southern "pine barrens"; and some of the ancient pine woods in New England were made up principally of this species. Such was that extensive wood near Concord, N. H., known by the poetic appellation of "Dark Plains," and in the early part of the century occupying a wide flat region in the valley of the Merrimack River.

This species does not give out its branches horizontally, nor in regular whorls. They run up at rather a wide angle with the stem, forming a head that approaches more nearly to a globular shape than that of any other of the American conifers. The branches have frequently a tortuous shape; for when crowded in a dense wood they do not so easily perish as those of the white pine, but turn in various directions to find light and space. They are likewise often bent downwards at their terminations, with a very apparent curvature. There is no conifer that displays so few straight lines in its composition; and, having no exact symmetry in its proportions, it may be mutilated to a considerable extent without losing its normal characters of beauty.

In young trees of this species the whorls of branches may be plainly distinguished; but as the tree increases in size, so many members of the whorl become abortive that all regularity of staging in their arrangement is destroyed. As these branches are numerous, with but little space between the original whorls, they seem to project from every part of the trunk. This tree displays very little primness in its shape, or of a spiry form, save when it is a very young tree. A peculiar habit of the Pitch Pine is that of producing little branchlets full of leaves along the stem from the root upwards, completely enveloping some of the principal boughs. These are rarely anything more than tufts of leaves standing out as if they had been grafted into the bark of the tree. It seems to be stimulated to produce this anomalous growth by the loss of its small branches. It then soon covers itself with this embroidery, and thus garlanded presents a picturesque appearance more interesting than that of the perfect trees.

I have seen very beautiful Pitch Pine trees of an abnormal shape, caused by the loss, when young, of the leading shoot. The lateral branches next below this terminal bud, being thus converted into leaders, produce two and sometimes three leading branches, giving the tree some of the characters of the deciduous species. The white pine is not improved by a similar accident, as it loses thereby the expression of grandeur that comes from the length and size of its lateral branches, which are always diminished by coming from two or more leading shafts. Michaux remarks that when Pitch Pines "grow in masses, the cones are dispersed singly over the branches, and they shed their seeds the first autumn after they mature. But on solitary trees the cones are collected in groups of four or five, or even a larger number, and will remain on the trees, closed, for several years."

The Pitch Pine abounds all along the coast from Massachusetts to the Carolinas; but it is rare in the northern parts of Maine and New Hampshire and north of these States. It is said to have been very abundant in the southern part of New England before the eighteenth century, but large forests of it were consumed in making tar for exportation to Great Britain. The Pitch Pine woods of the present day consist of small stunted trees, showing by their inferior thrift that they stand upon an exhausted soil.

The trees of this species, for the most part too homely and rough to please the sight, are not generally admired as objects in the landscape; but there is a variety in their shapes that makes amends for their want of comeliness and gives them a marked importance in scenery. We do not in general sufficiently estimate the value of homely objects among the scenes of nature, though they are indeed the groundwork of all charming scenery, and set off to advantage the beauty of more comely objects. They give rest and relief to the eye, after it has felt the stimulus of beautiful forms and colors, that would soon pall upon the sense; and they leave imagination free to dress the scene according to our own fancy.

Hence I am led to prize many a homely tree as possessing a high value, by exalting our susceptibility to beauty, and by relieving nature of that monotony which is so apparent when all the objects in a scene are beautiful. We see this monotony in all dressed grounds of considerable extent. We soon become weary of their ever-flowing lines of grace and elegance, and the harmonious blending of forms and colors introduced by art. This principle explains the difficulty of reading a whole volume written in verse. We soon weary of luxuries; and after strolling in grounds laid out in gaudy flower-beds and smooth shaven lawn, the tired eye rests with tranquil delight upon rude pastures bounded by loose stone-walls, and hills embroidered with ferns and covered with boulders.

The pines are not classed with deciduous trees, yet they shed their leaves in autumn with constant regularity. Late in October you may see the yellow or brown foliage, then ready to fall, surrounding the branches of the previous year's growth, forming a whorl of brown fringe, surmounted by a tuft of green leaves of the present year's growth. Their leaves always turn yellow before they fall. In the arbor-vitæ there is a curious intermixture of brown leaves with the green growth of the past summer; but, before November arrives, all the faded leaves drop, and the tree forms a mass of unmingled verdure.

FOREST CONSERVATORIES.

If we would preserve our forests, we must also, in about the same ratio, preserve the wild birds and animals that inhabit them. The woods are their houses; and nature has given them instincts and appetites that cannot be indulged, except by their performance of the very acts which are necessary to save their houses from destruction. While the woodpecker draws the larva from its cylindrical burrow, and while the bluebird seizes the beetle or the caterpillar that produces this larva, they preserve the trees in a sound and healthy condition by destroying those insects which, if they multiplied without any such check, would soon cause the entire forest to perish. But the birds, by their consumption of insects, are no more serviceable in the economy of the forest than by planting the seeds of the trees. As planters of the forest the small quadrupeds are as useful as the birds. There is not, indeed, an animal of any tribe, family, or species, which, if it be a consumer of fruit, is not also a planter of the vegetable that bears it. They all possess some kind of an instinct or habit that leads to this result. The jay and the squirrel, for example, constantly hoard nuts and grain, and, by hoarding, they plant that portion which they do not afterwards discover. The frugivorous birds distribute the seeds of all kinds of pulpy fruit as equally over the earth as it could be done by an artificial process. Even the grouse and the wild turkey are sowers of the grain that constitutes their food, by scratching a considerable portion of it into the soil.

It is a rational query, therefore, whether, if certain birds and quadrupeds were annihilated, the extinction of the trees and shrubs that bear a vital relation to them would not follow as an inevitable consequence, in the course of time. Those species whose seeds are scattered by the winds will never want a planter. But pulpy fruits and heavy nuts cannot be scattered by the winds, unless they be of destructive force; they can only be washed down a declivity. They cannot be carried upwards; nor, if the tree that bears them stands on a level, could they without the agency of certain animals be distributed much beyond the space covered by the tree that produced them. Yet if seeds were never strewn beyond the limits of the ground covered by the parent tree, the species, according to certain laws of vegetation perfectly well understood by the agriculturist, would ultimately become extinct. On this continent, which is destined to be crowded more densely with human population than any country on earth, it behooves us who are now living to take measures to prevent that sort of devastation which would cause it to be depopulated more rapidly than it was peopled.

I propose, therefore, that in every State of the Union, certain forest lands be purchased, comprising an area of three or four square miles each, in the ratio of one such tract to every square degree of latitude and longitude. These reservations should be kept as nearly as practicable in their primitive condition, or rather one closely resembling our primitive forest, by encouraging the growth of all their spontaneous productions, and excluding from them every description of culture. I do not mean that trees and other plants indigenous in each respective region, if wanting, should not be introduced, but that, after their introduction, they should be left entirely to nature. This tract should be a perpetual growth of wood, never to be disturbed for other purposes, to be set apart for the preservation of our indigenous plants, and the *fauna* of the region in which it may be located.

The ground selected for this purpose I would name a Forest Conservatory. It should consist of every variety of surface, upland and meadow, hill and dale, level and declivity, adapted to all kinds of native productions, and at the same time too barren and uneven to be easily reclaimable and reduced to tillage. A moderately barren tract should be preferred, if it be capable of sustaining a healthy growth of wood, because, on account of its inferior value, it could be more easily purchased. There are many diversified tracts of land in every region having a thin layer of soil, upon a rocky and uneven foundation, entirely useless for tillage, which would nevertheless sustain a heavy forest. By keeping any such place covered with wood we use it in the only way in which it could yield any particular advantage to the country. It should embrace within its limits either a pond or a part of a small river, for the benefit of water-birds; and a certain proportion of the soil should be tilled, for the purpose of sustaining the birds, which would otherwise be obliged to seek their living outside, and be exposed to the gunner.

The advantages afforded by these conservatories would rest upon the whole community, though only one person in a thousand should habitually frequent them. They would be seen in their effects on the local climate; in the growth of a magnificent wood in every section of country; in the preservation of interesting animals, that would otherwise be lost to that region, and in the multiplication of game, which would constantly overflow from this nursery into the surrounding country. These grounds would be mostly frequented by those who are addicted to the pursuits of natural history and the general study of science. Yet there are many who are neither scholars nor naturalists who would delight in an occasional stroll in these woods; the female sex, especially, among whom a certain refined intellectual culture is more general than among men, would find in them a charming resort, during a greater part of the year, under such guidance as would be provided.

The importance of forest conservatories, set apart from all purposes of commerce, agriculture, and manufactures, and designed to assist in regulating the beneficent action of nature over the whole wide country, cannot be easily appreciated. The masses do not understand the general principles upon which the meteorological effects of a regular distribution of forest are based. It is a subject almost as difficult as that of public finance. Our people know the value of timber in the mechanic arts, and of trees for shade, shelter, and especially for ornament; for a love of the ornate enters a people's mind before they have emerged from barbarism. But to understand the mysterious relations of trees and forests to the climate and soil, to drought and inundations, demands more study than men in general can devote to it.

I ought to remark in this place, that, when selecting a tract for a conservatory, the mountains must be excepted, because they are the resort of our people during the summer months. They are also the natural sporting-grounds of the nation, and the regulations needful to a forest conservatory would not be tolerated in those places, which have been adopted by the people for other and favorite purposes. The places that ought to be selected are not the mountainous districts, but unproductive spots of uneven and diversified surface, which would be completely desolated, like the bald hills of many parts of Essex County in Massachusetts, if all were left to the practical instinct of the community. These hills were formerly covered with a beautiful growth of timber. It would be less difficult now to cover them with soil by transporting it from the Mississippi Valley, than to restore their wood. The most barren hills may be kept in wood, after Nature has once planted them; but after their surface has been entirely cleared, their wood cannot be restored without the use of means which would be impracticable. The hilly and barren tract, still partially covered with forest, lying in Stoneham and Medford, called the "Five-Mile Wood," is well known to the inhabitants of Boston and its vicinity. A similar region I would select for a forest conservatory. Every State in the Union contains half-wooded barrens of a similar diversified surface, which might be dedicated to nature without diminishing in any degree the agricultural resources of the country.

But my present object is not to propose measures for the sole purpose of preserving our forests, but chiefly to afford asylums for the harmless creatures that originally inhabited them. This, if no other object were considered, has become a necessity, not only in the old States, but in every part of the country, to prevent that complete devastation which, in less than half a century, is sure to follow our rapidly increasing population, its love of field sports, and its material demands. Our defenceless animals must find a sanctuary in every region, or many species will be entirely extirpated from the States. I do not doubt that sufficient fragments of forest to harbor and sustain them would be left in all parts of the country; but what is to save them from the guns of sportsmen and of millions of boys who at all seasons of the year amuse themselves by shooting everything that has life? Some persons may not see the importance of preserving the lives of these creatures. It would be idle to reason with any one who should confess so much stupidity. Yet there is many an intelligent person whose attention needs to be called to the subject, before he can be assured that they are exposed to any such danger.

The life of every creature, except a few offensive and dangerous species, should be held sacred in this wood which is to be their sanctuary. It should be made a penal offence to kill a bird, or a squirrel, or any other harmless creature that should make its abode here. Many may believe that the wild animals need no such protection as these places would afford them, knowing that birds are as common in the densely peopled countries of Europe as in our own sparsely settled land. But we must remember that the whole land system of America differs from that of Europe, where many a nobleman owns a forest that is measured, not by acres, but by square miles, and of which his farm is only an appendage. In these princely forests the wild animals are protected by game-laws, which are the more easily enforced because they do not extend over the whole country. The game-laws enacted by our legislatures have proved inadequate to protect even the small birds of our orchards and gardens, and must ever be powerless for their preservation, on account of the difficulty of executing laws, over a wide extent of territory, which are opposed by the prejudices of the people. They might be enforced in grounds set apart for this purpose, but not effectually outside of them. Others having observed the great numbers of species that multiply in direct proportion as the land is tilled and cultivated and the wild woods removed, believe nothing is necessary for their protection. This subject I have explained in my essay on Animals of the Primitive Forest.

And what a rare opportunity would these grounds afford to those who would observe the ways of animals in their native habitats! The birds and quadrupeds would soon become tame and docile, from their constant security and familiarity with people who would not molest them.

Their habits, therefore, could be watched more carefully and minutely than in other wooded places, where they are terrified by the fowler, who causes them to look on man as their natural enemy. In a few years we should here witness scenes which have never before been exhibited on any magnificent scale. Visitors, not being allowed to shoot or entrap any living things in these grounds, would amuse themselves by making them their pets. It may be added that the habits of those who visit them would infuse a regard for the lives of these innocent creatures into the whole population, who would thenceforth afford them better protection in private grounds.

Here might be viewed assemblages of natural objects seldom found associated in other places; for though tillage is to be excluded, except so much as the object of the establishment requires, all the wild plants of every tribe and family, as well as the small animals indigenous to that region, would be gradually introduced. The student of natural history would find in this place many species now scattered widely and sparsely over the country, in localities which are difficult to be visited or discovered. He would be rewarded with the sight of species that have been absent from that region since the earliest clearing had been made of the primitive forest. Rare birds would sing there the first notes they had warbled in that region since the white man's cottage has taken the place of the Indian wigwam.

The expenses of such an establishment would not be great after the original purchase of the land, as the only improvements required or permitted would be the restoration of trees to certain vacant parts of the land, and the construction of a few paths to render the grounds conveniently accessible. These paths should not be gravelled. I would leave them to be overgrown by native herbs and grasses, and would take no further care of them, except to keep them clear of impediments. Nature would rear in these green wood-paths some of her loveliest productions; and the varieties of wild flowers that would spring up within them would form one of the principal minor attractions of the place.

I would discard every description of ornamental gardening from the conservatory, and I would admit no man as a superintendent who had been indoctrinated with any ideas of "æsthetics" as applied to the scenery of nature. Whenever it should become necessary to plant anything to fill a vacant space, or to supply some absent native species, it should be planted without any tasteful design. If any kind of embellishment were admitted, from that time the true object of the place would be forgotten. But I would not exclude the plough and other implements of the farm. Some amount of rustic tillage would be needful for the subsistence of the birds and animals which the place is designed to contain. But nothing should be raised for merchandise, and nothing cultivated for embellishment. This spot is to be nature's own sanctuary; no gravelled walks should disfigure it, no tasteful edifices should spoil its native features, no picnic-parties should desecrate it, no sportsmen should molest its inhabitants. No artificial seats or arbors should be provided for idlers or loungers. It should not be open to large parties, nor shut against any well-disposed individuals.

It might be supposed that an establishment of this character would need to be fenced, to protect it from the intrusion of persons who would not obey its regulations. A fence would not be necessary for this purpose. It would be sufficient to plant trees on the outside of the wood in such a manner as to render all entrance with a horse and carriage impossible. The most of the paths should be only a few feet wide, to accommodate pedestrians;

and the few driveways constructed should commence and terminate at the house of the superintendent, to be closed with a gate, and designed only for cartways and for the accommodation of invalids, not for recreation or amusement. A high fence, or a fence of any kind, would be useless. The beauty of the outside of a wood is greatly injured by a fence. The public should always enjoy the pleasure of beholding an unenclosed forest, which a person may enter on foot at any point and at any time.

By this proposal, I recommend no luxurious place of resort for wealthy men, who may pass their summers here in enjoyments from which others are excluded. A clause should be inserted into the charter of the institution that would render such diversion from the original object of the place an impossibility. The charter should be such that no man should be able to buy privileges which are not gratuitously offered to every individual of the community. We must consider that a place of this kind would offer no temptation to the classes who frequent fashionable resorts. As classes, neither the fashionable nor the vulgar would feel any special interest in it; but to certain individuals from every class of society these grounds would afford continual pleasure and recreation.

It may be objected that the time has not yet come for an enterprise of this kind; that the country is not yet sufficiently divested of wood to render such conservatories necessary for the purposes designed by them. It is admitted that we cannot pass over five miles of any road in New England without meeting with large fragments of wild wood, and assemblages of trees in some places of sufficient extent to be called forests. But these woods are liable at any time to be destroyed by the owner, who is not expected to preserve them a day after it is plainly for his interest to fell them. Whenever a tract of forest is cleared, the animals of all kinds that harbored there are expelled.

A few escape immediate destruction, by finding shelter in another wood. But, after being deprived of their native habitats, they are exposed to the fox, the hawk, and the gunner, and are soon destroyed. A gray squirrel is now as rare in our woods as the loon in our ice-ponds. Our wild animals are disappearing much more rapidly than our forests. Something must soon be done to save them, or in a few years none will be seen except in remote and unfrequented regions.

Let us now consider how such an establishment is to be supported. All similar places have been very expensive, for the plain reason that they are gardens, requiring the constant labor and attention of many hired men. This scheme involves no such expenses, for any description of costly labor would be fatal to its design. Here no idols of art are to be set up to divert men from the study of nature, and no costly appendages to be constructed for public amusement. But if an inhabitant of another State should visit Massachusetts to examine our native forest and its indigenous animals and plants, by entering one of these conservatories his curiosity would be gratified. He would there behold a forest inhabited by the same species of plants and animals that existed in that region three hundred years ago. Every State in the Union, containing a similar conservatory, would afford any stranger from another State or from abroad the same gratification. Under present circumstances, they must visit a great many different places, very remote from each other, to see in each what could be seen there in one natural assemblage.

As a superintendent of the grounds some person of superior knowledge should be appointed, who is sufficiently interested in the study of nature to be contented with a secluded life in this spot, and who would be satisfied with a moderate remuneration. What a delightful retreat would such a place afford to a clergyman whose pastoral cares had broken down his health, or to some enthusiast, who would gladly renounce the world to pursue his favorite science! These conservatories would offer so many frugal scholarships for the devotees of natural history, who would maintain an interesting correspondence with each other in all parts of the land, and whose observations would afford new acquisitions to science by which the whole world might be benefited.

INDEX.

A.

B.

C.

D.

E.

F.

G.

R.

S.

Cambridge: Electrotyped and Printed by Welch, Bigelow, & Co.

4741659

Made in the USA
Lexington, KY
25 February 2010